PENGUIN BOOKS

THE WASP THAT BRAINWASHED THE CATERPILLAR

Matt Simon is a science writer at *Wired* magazine, where he specializes in zoology, particularly of the bizarre variety. He is one of just a handful of humans to witness the fabled mating ritual of the axolotl salamander, as is detailed in this here book that he hopes you enjoy.

The Wasp That Brainwashed the Caterpillar

Evolution's Most Unbelievable
Solutions to Life's Biggest Problems

Matt Simon

ILLUSTRATED BY
Vladimir Stankovic

PENGUIN BOOKS

PENGUIN BOOKS
An imprint of Penguin Random House LLC
375 Hudson Street
New York, New York 10014
penguin.com

LIBRARY OF CONGRESS CATALOGING-IN-PUBLICATION DATA
Names: Simon, Matt.
Title: The wasp that brainwashed the caterpillar : evolution's most
  unbelievable solutions to life's biggest problems / Matt Simon.
Description: New York, New York : Penguin Books, 2016. | Includes
  bibliographical references.
Identifiers: LCCN 2016001823 (print) | LCCN 2016029575 (ebook) | ISBN
  9780143128687 | ISBN 9780698411258
Subjects: LCSH: Animals—Adaptation. | Predation (Biology) | Parasitism.
Classification: LCC QH546 .S58 2016 (print) | LCC QH546 (ebook) | DDC
  578.4/7—dc23
LC record available at https://lccn.loc.gov/2016001823

Printed in the United States of America
10 9 8 7 6 5 4 3 2 1

Set in Diverda Serif Com Light and Mixage ITC
Designed by Sabrina Bowers

For those earthworms I put on leaves and raced down flooded gutters when I was a kid. That wasn't funny. I was a jerk, and I'm sorry.

Oh, and my family. Them too. Not that I put them on leaves and raced them down flooded gutters. I mean I'm also dedicating the book to them.

Contents

Introduction

We need to talk about the wasps. I don't mean the little yellow and black things that menaced your childhood summers. Those are lambs, quite frankly. No, I mean, in no particular order: the one with a sting so powerful a scientist who has experienced it recommends lying down and screaming until the pain subsides, lest you run around in a panic and hurt yourself; the one that stings a cockroach in its brain and drags the zombie into a den, where the wasp's larva devours it alive; the one that opts instead to inject caterpillars with its young, which consume the hapless crawler alive from the inside out. Wasps are unparalleled in their ability to inflict suffering on other creatures, insects so cruel that Charles Darwin insisted a beneficent creator could never have thought them up.

But the thing is, in the animal kingdom, life sucks and then you die—as the saying goes. And out there, it's easy to die immediately. It's been that way for billions of years. For pretty much every creature (save humans), there's no slipping away peacefully in a comfy deathbed, because at any given moment some animal is trying to pull its head out of another animal's mouth. And I can guarantee you that something somewhere has a wasp larva consuming it from the inside out. Hell, a tree probably just fell on some kind of critter. *A tree.*

Nature is indifferent to death and suffering, and that's unsettling to us humans. We don't like thinking about an animal trying to pull its head out of another animal's mouth. There's no decency in that, for Pete's sake. But really, it's more than decent. It's beautiful. The predators and prey that grace this planet are the culmination of millennia after millennia of glorious evolution. From a single, ultrasimple organism all those years ago an explosion of life has radiated across the planet, and that life doesn't, well, always get along. And creatures don't have to worry about just each other: harsh climates and floods and tornadoes and asteroids are also cause for some concern.

Simply put, animals got problems. But at its core, evolution is the most majestic problem-solving force on planet Earth. Where it gets complicated is that it also *creates* all the problems. So things in the animal kingdom get a bit . . . involved.

Let's take, as an example, the plight of the zombie ant. It begins life as a normal ant in the rain forests of South America, foraging with its comrades along the colony's trails when, unbeknownst to the ant, it picks up a passenger: the spore of a fungus. Sticking to the ant's cuticle, the spore works its way into the host's body—and its mind. Here it releases chemicals that hijack the ant's brain, ordering it away from the colony and onto the underside of a leaf, always at a specific time of day at a specific height off the ground where the fungus can best grow. The parasite commands the ant to bite onto the vein of the leaf, then kills it and bursts out of the back of its head as a stalk to shower spores on the colony shuffling below. And thus the cycle repeats itself.

First of all, I didn't make that all up (we'll see the zombie ant in all its glory in chapter 4). And second, it's an unsettling illustration of nature creating and solving its own problems. At the base of it, to disperse its spores, a fungus would do well to have wind, which is lacking in the thick rain forest. So over millennia the fungus evolved a solution—use ants as vehicles. Yet the ants have their own solution to this problem: They instinctually grab individuals that look sick and drag them out of the colony and into a mass grave. But, alas, the fungus in turn has a solution to this problem: By manipulating the zombified ant out of the colony, it can avoid discovery. Thus one side evolves an offense and the other a defense, year after year, millennia after millennia. Push and pull, push and pull.

As if organisms didn't have enough to worry about with predators and even malicious fungi, the push and pull of problem and solution can bring conflict even between the sexes of a given species. You see, males and females don't have the same interests when it comes to sex. Males tend to take notice of anything that moves, while females have to be choosier. And then, randy fellas can come in conflict with other randy fellas. The males of one species of toad, for instance, have gone so far as to develop weaponized mustaches to battle each other for the right to mate. Even hermaphroditic species like some varieties of flatworm will clash among themselves, for when two individuals come together to mate, neither wants to get pregnant. Their solution? Penis fencing, obviously (coming up in the very first chapter, because I assume you're intrigued).

So what gives with all the conflict in the animal kingdom? Well, it's the system, man, the system. Specifically, Charles Darwin's idea of natural selection. Organisms must compete for food and water and often shelter, both with other species and with their own kind, and these individuals of course vary, due to errors during DNA replication and the unique way parents' genes mix for each offspring. Because there isn't always enough food to go around, not everyone is going to make it. If the ones that do make it have lucky genetics that help them win those resources, they can breed and pass down the primo genes, thus continuing the family line.

And food is just part of it. Those best equipped to escape predators, perhaps because they're that much faster than their peers, survive to pass down their genes. Those that can better tough it out in a harsh environment survive to pass down their genes. And those that are particularly impressive to the opposite sex, perhaps with exceptional feathers or dancing skills, win the right to pass down their genes. Conflict is everywhere, between predator and prey, brother and sister, sexed-up male and sexed-up female. A species may gain an edge, but any sort of edge is answered.

Weakness in the animal kingdom is dealt with accordingly, as creatures at times literally keep each other on their toes. Which is all to say that over the billions of years of life on Earth, evolution has created many a problem but also found many a solution. Push and pull, push and pull. And more often than not, things get really creative and really weird. This book is a journey through the strangest of the strange—a bestiary of sorts. And not a single one of these animals will die at the hands of a tree. A murderous fungus, maybe, but never a tree. You have my word on that.

The Wasp
That Brainwashed
the Caterpillar

CHAPTER 1

You Absolutely Must Get Laid

In Which Marsupials Hump Until They Go Blind and Die and Flatworms Stab Each Other with Their Penises

You're fond of sex, and that's okay. Everyone is—everything is—because living beings have to be. It's why we're on this planet: to pass down our genes to the next generation. You've got your tired pickup lines or that thing you do with your hair, or maybe, if you're feeling bold, both at the same time. And that's nothing to be ashamed of, because sex in the animal kingdom is far more ridiculous than anything you could possibly dream up. I'd guess, for instance, that you've never had so much sex that you died. Just a guess.

Antechinus

PROBLEM: The only reason any life is on this planet is to make babies. That turns out to be a lot of pressure.

SOLUTION: The males of a marsupial called antechinus mate with every lady they can find for three straight weeks until their hair falls out and they bleed internally and go blind and die, plus other bad things.

I know the meaning of life. I realize that's a rather bold thing to declare, but I really do know it. It is as follows.

Make a lot of whoopee. A *lot* of whoopee.

For the 3.8-billion-year history of life before we humans showed up and got all philosophical about our existence, critters on this planet had one goal and one goal only—to reproduce. Their secondary goals: eating enough to fuel the drive and not getting eaten to ensure the drive continued.

And there is no creature more committed to the cause than the Australian mouselike marsupial known as antechinus. Its males have so much sex, with so many partners, so consistently, that every single one eventually drops dead. But they're not perishing from something quick like a heart attack. No, no, that'd be too easy. This is total burnout, a burnout humans are incapable of suffering. As the males scamper about fornicating, they start to bleed internally. Their immune systems fail and their hair falls out. They even go blind toward the end, not that that's going to stop them. In a world that's faded to black, still they soldier on searching for females like sexual zombies, until at last they perish.

The issue is a whole lot of testosterone. Levels of the hormone skyrocket in antechinus males during the breeding season, which is great if you're looking to boost your sex drive, but not so great at promoting, oh, you know, emotional stability and general well-being. While on the upside, all that testosterone monkeys with antechinus's sugars so it doesn't need to feed for three weeks, allowing it to concentrate on sex sessions that last an admirable fourteen hours, it also leads to an unchecked release of cortisol, a stress hormone that supercharges energy levels but carries with it side effects like internal bleeding and hair loss and blindness.

So where are the females in all of this? Are they simply suffering these chumps, letting them run roughshod over the forest, humping willy-nilly? Well . . . yes, indeed, but they have far more control than it seems. In fact, over the course of the species' evolution, the females may have been responsible for creating the pandemonium in the first place.

YOU SAY POSSUM, I SAY *O*-POSSUM, BECAUSE I'M AN AMERICAN, DAMN IT

Australia is known as the continent of marsupials like antechinus, but you'll find plenty of marsupials in the Americas. (The sole species in the United States is the opossum, with an "o." Technically speaking, the possums without the "o" are a group in Australasia.) In fact, marsupials likely originated in the Americas, and at some point before 60 million years ago they made their way through Antarctica to Australia when the continents were still attached. Not that I'm trying to give the Americas credit for marsupials or anything. I'm just making a point.

Antechinus is insectivorous, and for an insectivore in Australia, nothing is more exhilarating than spring, when the populations of all manner of bugs explode. And it's in spring when the marsupial wants to raise her young because there's plenty of food crawling about, initially not so much for her kids to feast on, but for the mom herself. Marsupials like antechinus are born highly underdeveloped compared to other mammals, like horses, whose newborns hit the ground running (well, maybe more like stumbling, but still), and therefore must spend a whole lot of time drinking milk and growing. Mama antechinus puts a ton of energy into milk production for her young—which, by the way, sit not so much in a pouch like koalas or kangaroos as in a kind of bowl on her belly—and insects are her fuel. It seems that antechinus females have shortened their breeding season over evolutionary time to synchronize the weaning of their young off of the milk and onto solid foods with the very peak of insect abundance, helping guarantee their survival.

This in turn . . . guarantees the death of all those males the females mated with. Not directly, per se, but over the millennia males have had to adapt to solve the problem of a shorter breeding season by producing as much sperm as possible so they can shack up with as many females as possible as quickly as possible. Indeed, antechinus testes are enormous relative to their body size. By coupling with so many females, the males are making up for the abbreviated breeding season afforded to them by evolution.

It might all sound counterintuitive to the survival of a species, having males and females locked in a kind of evolutionary antagonism. But really it's the opposite. An antechinus female is just asking more of the males. Yes, she's mating with a bunch of fellas over those three weeks, so she can't be choosy about her partners, unlike, say, a peahen, which can select the peacock with the most magnificent plumage. But because the healthiest males produce the most sperm, they have a better shot at fertilizing her. In her own way she's "selecting" for the best genes of the bunch. Plus, her brood may be made up of young fathered by multiple males, and

ADVENTURES IN HAVING A SIX-INCH-LONG CLITORIS

Hyenas employ a more active approach to selecting the sperm of certain males after copulation. Females have six-inch-long clitorises that look like penises. When they mate, the male inserts his penis into her "penis," an act that biologists have noted takes some practice on the male's part. The female's enormous clitoris may have evolved to allow her to urinate to flush the sperm out if she decides she isn't all that into the male after all. Perhaps unsurprisingly, throughout history all of this has earned the hyena a reputation as a sexual deviant. None other than Ernest Hemingway once described a hyena named Fisi as a "hermaphroditic, self-eating devourer of the dead, trailer of calving cows, ham-stringer, potential biter-off of your face at night while you slept."

Boy, that Ernest. Always riled up about something.

when she gives birth to as many as three times more babies than she has teats, only the strongest will win the struggle for nipples. The rest will perish and take their fathers' inferior genes with them.

To our eyes such scenes are brutal, but this is how life works, a rather grim illustration of Charles Darwin's 1859 theory of evolution by natural selection. That *by natural selection* bit is important, because naturalists in Darwin's time had already been kicking around the idea that species change, though they usually referred to this as "transmutation." Darwin's society-shaking revelation was the mechanisms involved: Species typically produce more offspring than can survive, these offspring vary in their characteristics, and those with variations that better suit them to their environment survive to breed and pass those "good" genes along. This is how a species evolves, adapting to its environment and predators. And necessarily a whole lot of maladapted creatures must die in the process.

Antechinus blokes, for their part, don't seem to mind it. They're simply living out the meaning of life: get laid at all costs, even at the expense of their own well-being. And really, females live for only a few years, so your regular Joe Schmo Antechinus doesn't have much to lose with his suicidal plunge into fatherhood (not a single one of them makes it past his first birthday, since they were all born after the previous mating season). He's mated with dozens of females, and ideally at least one of them will bear his young. Meaning of life achieved, he slips into darkness as the females prepare to do the real work.

Anglerfish

PROBLEM: Think finding love in a bar is hard? Try finding it in the desolation of the deep sea.

SOLUTION: When the diminutive male anglerfish finds a lady—who can be five hundred thousand times heavier than him—he doesn't let her go. He bites onto her, fuses his face to her tissue, and lives the rest of his life releasing sperm whenever she beckons.

Let us not mourn poor antechinus, for there's an hombre down in the ocean depths that doesn't have it so easy as death by sex. This one lives his life in sexual servitude, and not even the good kind. I'm talking about the freakish, not-at-all-fun existence of the deep-sea anglerfish.

You'd be forgiven for thinking that the females and males of the 160 or so species of deep-sea anglerfish were different animals. The females are stunning. First and foremost, they have a bioluminescent lure that dangles in front of their faces to attract prey as well as mates, with species utilizing different lure shapes and flashing routines so the roaming males of their respective kind can tell them apart. They've got giant teeth attached to giant mouths so they can be sure to consume whatever rare prey they happen upon in the desolation of the deep, as well as cavernous stomachs that allow them to digest such meals. And indeed the many species of anglerfish differ quite remarkably in their shapes and sizes. Some are more streamlined, while others are comically spherical stomachs with faces, forsaking all that "speed" and "maneuverability" for the life of a swimming softball.

Sometimes female anglerfish have tiny lumps on their bodies that appear to be parasites latched on for a free meal. These are parasites, all right, in their own way, at least: They're actually small male anglerfish, so small that in some species the fellas are among

LIGHT OF MY LIFE

Perhaps 90 percent of life in the deep oceans utilizes bioluminescence. This would seem to be somewhat of a detriment to survival, what with being lit up like the Fourth of July down there in the dark for predators to see. But bioluminescence is an indispensable tool for a number of reasons. Beyond its usefulness for communication and luring, as we see in the anglerfish, creatures can also use it as a defense system: Some species of shrimp eject a cloud of glowing goo to confuse attackers like an octopus ejects ink. Others use bioluminescence to light up their bellies to match what little light is coming from the surface, breaking up their silhouette for predators watching below. And still others employ what is known as a sacrificial tag: If a predator eats part of their glowing body, and that predator is transparent (which happens to be in vogue down there), they'll keep on glowing in its stomach. This makes the predator suddenly the conspicuous prey, like a bank robber marked by an ink bomb.

the world's smallest vertebrates (that is, creatures with a spine, as opposed to invertebrates, which lack a spine). These lucky males have fulfilled what is their only function in life: to find a female, which can weigh five hundred thousand times more than them. In fact, males are unable to feed, and they set out from birth to find the elusive females down in the darkness. Yet only an estimated 1 percent of males ever succeed. The rest die of starvation. But should they be able to sniff out a female—using the biggest nostrils relative to head size in the animal kingdom—they'll latch onto a lady with their pincerlike teeth, and the pair will live out the rest of their lives as one.

Here's how it works. After the male has latched on, enzymes begin to break down his face, welding him to the female. He'll grow considerably larger by tapping into the female's circulatory system, which will provide him with nutrition, while unnecessary organs and structures like the eyes atrophy away. Because he's stealing nutrition from her, he's technically a parasite, and in fact this is known as sexual parasitism. That's right, anglerfish are parasites of . . . themselves. (The males of some species, though, are fortunate enough to forgo the fusing: They attach, take nutrients from the female and release some sperm, then detach and go on their way.)

The charming couple also syncs hormones, so the female can trigger his sperm release when she releases her own eggs, great thirty-foot-long sheets of them that soak up his seed. (Because the male is providing something to the female—that is, sperm—some biologists might argue that he isn't technically a parasite. So the anglerfish relationship is complicated, to say the least.) The female has become, in effect, a self-fertilizing hermaphrodite. Over her thirty-year lifetime, she may collect several of these males, all of which keep producing sperm until she dies and the whole weird amalgam sinks down into the abyss.

This rather inventive solution to the problem of finding regular sex has the added benefit of turning the anglerfish into a baby-making machine, churning out as many eggs and sperm as her energy levels allow. That's why female anglerfish so dwarf their male counterparts: It's all about the gametes—the eggs and sperm. Human males tend to think that their size grants them some sort of superiority over women, and that the same must be universally true in nature, but in most species the females actually grow larger than the males. That's in part because females must put a ton of energy into producing eggs—the lady anglerfish has such a giant mouth and stomach because she lives where there are so few things to eat, and therefore needs to be sure to tackle whatever comes her way to procure the energy. Eggs also take up far more room than sperm, so the male can afford to be comparatively tiny.

Elsewhere in the animal kingdom, eggs are so precious that females are quite picky about who they mate with because they can only produce so many gametes. They shack up with only the best dancer or scrapper, for instance, to be sure they're choosing strong genes to pass down to their kids. And when animals like mammals do get knocked up, they have to put even more energy into raising the little rug rats inside their bodies. Such pickiness leads to intense competition between males for the right to mate, which can result in some species flipping the size norm: Males grow bigger than females in order to compete among themselves for the ladies' affection.

The female antechinus and anglerfish, though, have evolved to "choose" the best-adapted males in their own subtle way. The former makes her counterparts grow big testicles while the latter is very elusive—remember that all but 1 percent of male anglerfish perish before finding a mate. But that's not for lack of trying. It's just really, really hard to track down a female in the vast blackness of the deep. Only the best, master-lady-tracker males have the chance of fusing to females and passing along their genes to the next generation. This is probably why over evolutionary time the males' mate-sniffing nostrils grew so big, for such a trait wins them the right to pass down their genes for a large schnoz.

It's no wonder that these weird goings-on confused the first naturalists to observe attached males in the early twentieth century, scientists who originally thought they were seeing larvae fused to the females. But when the truth emerged, confusion turned to shock, which then gave way to horror. In 1938, William Beebe, the great naturalist and explorer, summed up the anglerfish zeitgeist of the day best:

> To be driven by impelling odor headlong upon a mate so gigantic, in such immense and forbidding darkness, and willfully to eat a hole in her soft side, to feel the gradually increasing transfusion of her blood through one's veins, to lose everything that marked one as other than a worm, to become a brainless, senseless thing that was a fish—this is sheer fiction, beyond all belief unless we have seen the proof of it.

So you keep swimming around furnishing that proof, little anglerfish sir. And you keep being you.

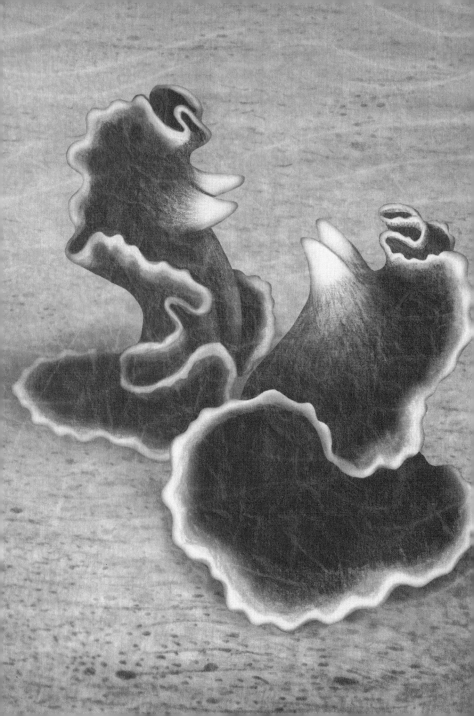

Flatworm

PROBLEM: Being a mom is a huge responsibility.

SOLUTION: Hermaphroditic flatworms penis-fence, of course. Individual flatworms have both sperm and eggs, so whichever worm stabs its partner with its needle-like genitals wins the honor of not giving birth, while the fertilized loser mopes away.

As we've seen with the anglerfish and antechinus, males typically have to put in the effort to get laid, but the true biological burden lies with the female, who expends enormous energy producing eggs and, in the case of mammals, bearing and looking after the young. But what if you're a hermaphrodite, like some species of marine flatworms? Who's going to bear the maternal burden there?

Whoever loses the bout of penis fencing, that's who. In coral reefs, far away from the depths where anglerfish have their own weird sex, certain species of flatworm do battle with their members.

It starts off innocently enough. Two often brilliantly colored flatworms approach each other and nuzzle a bit. But before long the calm departs, as each rears up and exposes its weapons: the two sharp white stylets that are its penises. Like human fencers, each flatworm will juke and stab, simultaneously trying to inject its partner with sperm anywhere on its body while doing its best to avoid getting inseminated itself. And this can go on for as long as an hour until the two retract their double penises, lower themselves, and go their separate ways. When the struggle is over, both

ANYTHING CAN BE A WORM IF IT PUTS ITS MIND TO IT

As far as the English language is concerned, a worm can be just about anything, within reason, of course. If you're elongated and don't have a spine and are kinda squishy, chances are you can pass as a worm. So this includes the many-legged velvet worms (we'll meet them later), the tapeworms that invade our intestines, and the earthworms, which in Australia grow to six feet long (leave it to Australia, the Land of Generally Ridiculous Beasts). That length is laughable, though, compared to the marine bootlace worm, which has been known to grow to one hundred feet long—the length of a blue whale, the biggest creature ever to grace planet Earth. One specimen found in Scotland was even said to measure in at 180 feet, though its handlers—perhaps chasing that ever-elusive title of "Discoverers of the Longest Creature on Earth"—likely stretched it out. Fame does weird things to people, you know.

can end up pockmarked with white stab wounds filled with sperm, and you can see pale streaks running along their bodies, branching rivers of semen on their way to fertilize the eggs.

Now, you might be asking *why*. Why evolve violent "traumatic insemination," or more specifically and hilariously, "intradermal hypodermic insemination"? The problem is that the two flatworms have the same interest: Neither wants to be the female (I know that sounds sexist, but bear with me here). Developing those eggs is a tremendous energy suck, not to mention that the loser is deeply wounded on top of being knocked up. The winner gets to pass down its genes without taking the trouble of raising the young.

But here's the weirdest part. Natural selection dictates that if the tapeworm's going to get stabbed, it's in its best interest to get stabbed a whole lot. The most accomplished fencers are the ones who will have the most reproductive success, and their genes are what other flatworms want to pass down to their offspring, who will in turn be more likely to become skillful combatants and fertilizers. It's one of nature's cruelest ironies: The flatworm doesn't want to be stabbed with a penis and inseminated, but if it must, it may as well get stabbed with a penis and inseminated thoroughly.

Things get even weirder with another type of flatworm, this one a transparent, microscopic species. It, like our beautiful seafloor variety of flatworm, mates by injecting its partner with sperm. But it seems that the tiny flatworm can really feel the pangs of loneliness: If there aren't any partners around, it uses its stylet to stab *itself* . . . in the *head*, a maneuver known as selfing. The stylet is at the tail, while the head is of course at the other end, so with a dexterous bend the flatworm can jab itself right in the noodle. The sperm then makes its way down the body to fertilize the eggs. So in a pinch, the flatworm can reproduce all on its own. The researchers who discovered the behavior cautiously referred to it as *hypodermic* insemination, not *traumatic* insemination (as in the aforementioned fencers), because they weren't sure if the creature seriously injures itself with the stab to the head. Not even kidding here.

Now, flatworms aren't the only creatures out there engaging in such shenanigans. Far from it. In case you needed another reason to fear/despise/be grossed out by bedbugs, they're reproducing by traumatic insemination in our sheets. A male will puncture a female's exoskeleton with his genitalia and pump his sperm into her body cavity—no trifling matter when bedbugs rely on their tough shells to protect them from the elements. Indeed, female bedbugs have evolved an immune response: proteins that erode the cell walls of bacteria, helping them ward off infection.

Such is the push and pull in the battle of the sexes. As one side evolves an attack, the other evolves a defense, nature creating problems and then solving them. The issue comes right back down to the meaning of life: reproduce at all costs. This can put the sexes in conflict with each other—or, in the case of the tapeworm, the single hermaphroditic sex in conflict with others of the single hermaphroditic sex—particularly when females need to maintain some measure of control over who they mate with to ensure they're picking the best genes. And perhaps nowhere is this kind of sexual conflict more dramatic than among ducks, whose males are notoriously forceful with their mating. Females have evolved a vagina that corkscrews to try to keep out the male's penis, which corkscrews in the opposite direction (and can grow up to fifteen inches long). Some duck vaginas even have pockets that branch off into dead ends, frustrating the male's efforts.

The idea that animals can be choosy about their mates, and that such choosiness will drive the evolution of certain characteristics, was one of Charles Darwin's more brilliant realizations. Known as sexual selection, it drew ridicule in Victorian England, a patriarchal society that found the notion of female choice laughable, to say the least, especially when it came to sex. A notable dissenter, though, was none other than Alfred Russel "So What If It's Only One 'L'" Wallace, the phenom naturalist who had simultaneously developed the theory of natural selection on his own. (Charles Darwin scrambled to publish *On the Origin of Species* after receiving a letter from Wallace, an acquaintance who would later become a good friend, pontificating on his ideas. But that was only after colleagues presented the ideas of both men to the Linnean Society of London.) Wallace didn't think animals had the brainpower to make these choices, except when it came to the ladies of our own species. He wrote that "when women are economically and socially free to choose, numbers of the worst men among all classes who now readily obtain wives *will be almost universally rejected*," thus improving the species. Emphasis his own.

Gotta love that feminist optimism, however wrong he may have been about sexual selection. (To be clear, Wallace was brilliant, so perhaps this isn't the greatest way to introduce him, and

THAT TIME WHEN FOX NEWS AND DUCK GENITALS COLLIDED—NOT LITERALLY, OF COURSE

One scientist who studies the duck genital arms race, Patricia Brennan, came under fire in 2013 when conservatives discovered that she received federal funding for her research, through the National Science Foundation. It seems that at the time a Fox News poll found that 89 percent of its website readers/TV viewers considered the study of ducks' naughty bits to be a waste of their tax dollars. Brennan defended herself in a *Slate* op-ed, noting that it was actually during the Bush administration in 2005 that the government first gave her money. "The fact that this grant was funded," she wrote, "after the careful scrutiny of many scientists and NSF administrators, reflects the fact that this research is grounded in solid theory and that the project was viewed as having the potential to move science forward (and it has), as well as fascinate and engage the public."

Duck genitals: Fascinating Americans, other than 89 percent of Fox News viewers, since 2005.

for that I apologize. But we're all wrong sometimes, and indeed being wrong is fine in science, for it invites others to discover the truth. We'll see more of Wallace in the next chapter being very right.) Females in the animal kingdom can in fact wield great power when it comes to sex.

So, sure, we human men may not always have the greatest ideas, but at least we're not penis-fencing. It's the little things that count, really.

Mustache Toad

PROBLEM: Competition to win the affection of females can be intense.

SOLUTION: If you're a certain kind of male toad, you grow a mustache with which to shank your rivals.

"The sight of a feather in a peacock's tail, whenever I gaze at it, makes me sick," Charles Darwin once wrote to the American botanist Asa Gray. He despised the peacock's flamboyance, for it seemed to be a living, prancing affront to his theory of evolution by natural selection. He agonized over how something so heavy and conspicuous could make a creature anything more than vulnerable. But Darwin eventually realized the tail's ability to attract females outweighs its risk of getting the owner killed: sexual selection. The peacock is a tasty creature with a giant target sprouting out of its bum, but it's also a *sexy* creature with a giant target sprouting out of its bum.

But for other horny fellas out there, such foppishness seems a luxury. Sometimes the right to mate is won not with towering feathers but with much bloodshed and broken bones and maybe even a ruptured organ or two. A male's mad pursuit of passing along his genes sometimes overrides the risk of bodily injury.

There are few creatures that go about it all more fashionably than the mustache toad, whose name speaks for itself. The thing is horny. Just as male deer shed and regrow their horns to battle each other every mating season, so, too, does the toad develop its own weapons: ten to sixteen extremely sharp spikes situated along its upper lip. Known as nuptial spines, these lances are made of keratin, the same stuff that makes up your hair and fingernails. And they grow right through the toad's skin. With these spines the toads engage in hyperviolent battle with other males for prime mating spots.

Now, these toads are terrestrial creatures, but each year they return to the water to mate in an event epically known as "explosive breeding." Competition to mate is daunting, as toads from all around begin gathering for their three-week sexy season, with ideal mating spots being underwater. If you were a smart mustache toad you'd be looking for a nook with an overhang, a spot where you can best guard your developing young. This makes breathing a problem, since coming up for air is risky when there are plenty of other gentlemen that would steal a good base of operations. But the toad has an ingenious, flappy solution: Its skin gets loose and wrinkled in the mating season. That may seem like a glaring inconvenience and certainly not a solution at face value, but it boosts the creature's surface area. And because toads can absorb oxygen right out of the water, more skin means better breathing and less frequent trips to the surface.

Inevitably, though, the squatters will face challengers. The battle begins with some good old posturing, as the defender tries to block the entrance to the den while vocalizing. But the challenger's having none of it. He attacks, and the two come together like flabby-skinned sumo wrestlers, grappling and shoving and squirming. The scrappers start stabbing, both trying to get their snouts under their opponent so they might heft each other up. If one of them succeeds, he'll lift his foe vertically, still holding tight, and proceed to shank him in the belly while slamming him against the walls of the den.

While the foes may have loosened skin, it does little to protect them from the trauma. Getting laid is great, but getting killed in the process doesn't help the cause, so the weaker toad flees, exhausted and poked full of holes. In the end almost 100 percent of the combatants will end up injured in some way. But the stakes are

ALL RIGHT, WHICH ONE OF YOU KNUCKLEHEADS NAMED THE IRISH ELK?

The mustache toad may have great facial hair, but its weapons are like Tinkertoys compared to those of the Irish elk, which went extinct just eleven thousand years ago. The animal wasn't confined to Ireland and was actually a regular deer, not an elk, but it did have the biggest antlers of any animal ever to live. With antlers twelve feet wide, you can only imagine the toll these things took on their owners, and in fact that may have helped do the awe-inspiring Irish elk in. As their habitat began to cool, the food supply dwindled, a problem of particular severity if you're in possession of twelve-foot-wide horns that you need to feed in order to regrow every year.

too high not to fight for a cave: Females are sizing up not just a mate, but his den as well, since he'll be taking care of the kids while she absconds. When she comes along, there's a bit of nudging with the victor, and a bit of swimming around checking the place out, and finally the sexy time. He grasps her, and while she lays eggs, he fertilizes them and sticks them to the roof of the den with his foot. She then splits, while the male remains, rubbing up against the eggs to keep them clean until they hatch and swim off as tadpoles.

If they haven't snapped off in combat, the nuptial spines will fall off on their own as the toad returns to land and a life of relative

APPARENTLY NECROPHILIA IS "SOCIALLY UNACCEPTABLE"

The females of another species of frog, *Rhinella proboscidea*, don't get off so easy when it comes to mating. These frogs are also explosive breeders, with hundreds amassing in a single pond. In the frenzy, many females will drown, but that's not about to stop the males. They'll massage a dead female's belly to eject her eggs, then fertilize them. It's called functional necrophilia, because why not. In one paper on this bizarre ritual, scientists found it necessary for whatever reason to note that this ranks among the reproductive behaviors "that are considered 'socially unacceptable' and impossible in human society."

peace. And I really can't overstate what an amazing commitment the whole adventure is for the males. On top of holding their breath and hanging out in caves for the three weeks of the breeding season, growing and shedding spines out of their faces year after year is a huge drain on resources and energy. But again, the stakes are too high to not participate in the strange rumpus.

At the very foundation of all this is the raw fuel of evolution: variation. When two parents come together, their offspring vary not only because their two sets of genes combine in unique ways for each kid (save for identical twins, of course), but because mutations can sneak in. These mutations can be either harmful, inconsequential, or beneficial to the organism. So a mustache toad isn't just an exact copy of the previous generation—at some point long ago, male toads with mutations that gave them mustaches started showing up, and because that trait afforded them a competitive advantage to win mates, they got to pass along the genes responsible for the spines. And winning mates is only one consideration here. Traits that help organisms eat in turn help them survive to pass their genes along. Traits that help them avoid predators do the same. Thus species evolve to fit their environment, skirt their enemies, and earn the right to mate.

And thus things get a bit carried away sometimes, resulting in adaptations like weaponized mustaches. But there's no guiding hand here, just a step-by-step transformation of an ordinary toad into an amphibian Tom Selleck.

Toadfish

PROBLEM: As with the mustache toad, fish sometimes have to win the right to mate.

SOLUTION: Fighting is for simpletons. Instead of brawling, toadfish males use their swim bladders to attract females. They produce a hum so loud that the mysterious noise once drove houseboat residents in an American town to madness and led to a full-scale investigation.

The year was 1985. The locale: a seaside community called Sausalito just north of San Francisco, where houseboats bob and sometimes a sea lion shows up, much to the delight of tourists. The problem: a maddening, perhaps excruciating hum on summer nights that some said was the work of aliens. Others fingered the government, and still others the nearby sewage plant. The curious editorial headline to mark the occasion in the *Marin Independent Journal*: "Do lovesick fish sing in Sausalito?" The paper's answer was an emphatic no.

But the paper was wrong. The town had in fact fallen under the spell of a mating song—the toadfish's, to be exact—a sound that defies human credulity. It's reminiscent of a bassoon, or an extremely loud swarm of bees. An official complaint from a houseboater described it as a noise "similar to having an airplane in your house," which in retrospect may have been a bit hyperbolic. The clamor is a bummer for humans, sure, but really the male toadfish has no choice in the matter. It needs to attract females, who choose only the boys with the most impressive bellows.

WHO ARE YOU CALLING A LYRE?

No offense to the toadfish, but when it comes to mating songs, nothing on Earth can beat that of the lyrebird. When vying for the affections of the ladies, the male will pop his resplendent, almost peacocklike tail feathers over his head and prance around, all while belting an astonishing tune. It's part sci-fi laser, part high-pitched plucking of strings, and part impressions of other birds in the forest. In captivity, though, he'll imitate the surrounding cacophony: car alarms, drills, and hammers (if he's in a zoo that's undergoing renovations), even camera shutters. It's all done so perfectly that it boggles the mind. At the risk of tearing you away from this here book, you should go find videos of the lyrebird. I mean it. It won't hurt my feelings.

The word spilled out of Sausalito and went national. On the *CBS Evening News*, Dan Rather was in a state of disbelief, declaring that if you think that a fish could produce such a din, "you believe frogs have hair." (It turns out there is indeed a species called the hairy frog, whose males have hairlike projections of skin that, like the mustache toad's loose skin, help them breathe underwater while guarding eggs—but whatever.) The *Marin Independent Journal*'s editorial was similarly unrestrained and seems to have been the work of either satirists or the functionally insane, or perhaps both. The paper noted the timeliness of the noise—generally from nine p.m. to five a.m.—and asked, "What self-respecting fish keeps a schedule?" They added, no joke: "Only the species that developed the atomic clock, the digital watch and the on-time European train is capable of creating such a commotion at the same time night after night."

The man who set off this hullabaloo was one John McCosker, a legendary biologist at the California Academy of Sciences in San Francisco. *So*, he tells me, one day back in the eighties he got a call from a noise specialist in the health department, which had been investigating Sausalito's racket and spending a good chunk of change in the process. They'd run out of viable theories, and the last and least likely on their list pegged the hum as biological in origin. McCosker had the guy play a recording over the phone, and without hesitation he identified the racket as the product of *Porichthys notatus*, the toadfish.

"Jesus Christ," McCosker recalls the man saying. "*Don't. Tell. Anybody.* Because we've been going all the wrong places trying to figure out whether it was the sewage plant or the Army Corps, and all of these people are *really* pissed off about it. I didn't know a fish could make so much noise."

The good people of Sausalito weren't buying it. What they had a hard time coming to terms with is that the ocean is an incredibly noisy place to be because sound travels very well in water. Humpback whales, for instance, sing thundering songs that last as long as a half hour, communicating with each other across entire oceans. All manner of fish and porpoises are breaching, slapping the surface of the sea. The waters themselves are in an uproar, as tides and currents and upwellings from the depths clash, even far from shore. As marine biologist Rachel Carson once described it in her classic book *The Sea Around Us*: "Superficial hissings and sighings, the striping of the surface waters with lines of froth, a confused turbulence and boiling, and even sounds like distant breakers accompany the displacement of the surface layers by deep water." Thus water and creatures conspire to produce cacophony. Perhaps it was McCosker who put it best in a 1986 paper: "It is my view that the Bay Area yuppie invasion comprises a generation that has all but forgotten the noises of nature—like crickets, frogs, cicadas, and toadfish. Damn shame."

The toadfish male's trick is a gas-filled organ called a swim bladder, which helps fish maintain neutral buoyancy. By control-

THEY DON'T MAKE CONDOMS LIKE THEY USED TO, AND THAT'S FANTASTIC

Fish bladders aren't useful just for fish. European gentlemen once used them as condoms. From a 1908 sales catalog: "Fish bladders are preferable to rubber since, being significantly finer and more durable, they are not as obtrusive as rubber, and the sensation is hardly affected if at all." These things were really expensive, being the ultrasensitive rubbers of their day, and accordingly conscientious gentlemen would wash them after each tryst and reuse them. So it would seem that while the swim bladder has helped the toadfish make babies, it's also helped humans avoid making them.

ling how much oxygen fills its swim bladder, the toadfish can avoid constantly adjusting up or down with flips of its fins, thus saving energy. (Sharks, famous for energy conservation, use a similar method, but they don't have a swim bladder. Instead, they maintain their position with the help of their giant liver, which may make up as much as 25 percent of their weight.) The male toadfish can vibrate the muscles that attach to the resonating chamber as many as 150 times a second, producing a hum that drives the ladies—and humans in the houseboats above it—mad.

Yet not all male toadfish hum. The males are split into two types: an alpha variety, which actively hums for females, and a so-called sneaker variety. The alphas are eight times larger than the sneakers and have swim-bladder muscles that are six times as big. The sneakers, though, have the alphas beat on testicle size, sporting gonads that are seven times as large as those of their counterparts.

But why? Well, the alphas maintain nests on the seafloor, and they sing and sing and sing, for perhaps as long as two hours straight, with those big, highly developed muscles powering their swim bladders. The females they manage to woo will approach and drop off a batch of eggs, and the alpha will fertilize them and continue to look after the developing young. But a sneaker male will lie silently in wait nearby, then dash in and fertilize the eggs himself. And it's here where those big testicles probably come into play: The sneaker may get only one chance at fertilizing a clutch, so he wants to produce as much sperm as possible in a go. The alphas, by contrast, invest more energy and resources into growing giant to defend their nests. But the sneaker avoids devoting energy to all of that stuff—growing outsized, building a nest, calling to the females, and looking after the eggs once they're fertilized. Thus the genes that make for a clever sneaker toadfish persist through evolutionary time, however precariously alongside the pissed-off alphas that are doing the actual work.

And evolutionary time is a whole lot of time, so why did the toadfish's love song suddenly become a problem in Sausalito in the eighties? Where had they been all along? Well, they were there, all right, but in plummeting numbers. During World War II, a nearby shipyard was churning out war machines, and lots and lots of nasty chemicals along with them. The bay was a mess, and the fish suffered for it. But then cleanup crews went to work, McCosker says, "and did some dredging and stopped putting all of the spoils in there, and more fish starting coming in." The Sausalito yuppies, and their admirable environmentalism, had ironically enough been responsible for the racket all along.

After McCosker's revelation, disbelief gave way to acceptance, and for a while the residents of Sausalito held a toadfish festival. That's gone now, but the toadfish remain, humming their love song as the houseboaters above wonder how big a war it would take to get that shipyard rebuilt.

CHAPTER 2

You Can't Find a Babysitter

In Which Caterpillars Give Horrific Birth to Maggots While Other Caterpillars Give Stylish Birth to Hairdos

Thank your lucky stars you're a mammal, and thank them again that you're a primate whose mother could cradle you and cart you around. We humans look after our young, but few other animals have such luxury. So in the absence of babysitters, they rely on other strategies to ensure their kids survive. Some just throw a lot of offspring into the ecosystem and hope a few make it, while others get rather more creative with the whole thing. As in, injecting their young into other animals and passing off the responsibility. So, again, thank your lucky stars.

Ant-Decapitating Fly

PROBLEM: Maggots are helpless. And that can be an issue if you're a maggot.

SOLUTION: The ant-decapitating fly surgically inserts its kid into a living ant, where the maggot moves into the brain and mind-controls the host into the leaf litter, before releasing a chemical that pops the ant's head off. Safely inside, the maggot develops like a babe in a crib.

If evolution by natural selection is the greatest idea anyone has ever had, its discovery is surely one of the more stunning coincidences in the history of human thought. Good Old Charley Darwin gets all the credit, but the aforementioned Alfred Russel Wallace had not only simultaneously developed the same theory, but had arrived at it, at least in part, by the same inspiration: Thomas Malthus's "Essay on the Principle of Population." Malthus argued that if the human race didn't keep its population in check, it'd have the checks placed for it in the form of war and famine and intense competition for resources. Both Darwin and Wallace realized that the same is going on in nature: Species have more offspring than can survive, and predation and limited food pare them down and keep them in check. Siblings vary, so the ones with the beneficial variations survive and pass down their genes, driving evolution.

PUTTING DEAD SNAKES IN LIQUOR AND OTHER BORNEAN PARTY TRICKS

Darwin and Wallace may have stumbled upon the same superb idea, and they may have been good friends despite Darwin taking the limelight, but the two men couldn't have lived more different lives if they'd tried. Darwin came from wealthy stock, which allowed him to study and write without worrying about making a living, while Wallace had no such privilege.

The voyages that led them to their discoveries couldn't have been more different either. Compared to Wallace, Darwin had a downright comfortable trip around the world aboard the *Beagle*, finding danger and toils all along, sure, but also good lodgings among wealthy expat Europeans while wandering on land. Wallace—a lanky, bespectacled man—rotted in the jungles of Southeast Asia, earning what you'd generously call a living by shipping specimens back to European collectors.

Even preserving the specimens in the first place was a tall order for Wallace. On Borneo, he pickled them in a local liquor called arrack, which unsurprisingly was a favorite of the local people. "To prevent the natives from drinking it," Wallace writes in *The Malay Archipelago*, a fantastic account of his travels, "I let several of them see me put in a number of snakes and lizards; but I rather think this did not prevent them from tasting it." Whether it ever crossed his mind to give up in despair and just drink all the liquor himself is sadly lost to history.

When it comes to explosive populations to rival humanity, there are few species more successful than the ants. Conservatively estimated, they number 10,000 *trillion*. But as the old adage goes, haters gonna hate, and there are a *lot* of ant haters out there willing to keep the ants in check. Of them, perhaps the most creative and brutal is the ant-decapitating fly. Yes, that's what it's called. And yes, that's exactly what it does: decapitate ants . . . from the inside out.

Getting laid is all well and good, but it'd be nice to be able to protect your young as they develop, since you don't want that fancy jousting and humming and having sex until you die to be for naught. So instead of leaving its larvae to the mercy of predators and the elements, the ant-decapitating fly goes out in search of babysitters, specifically the fire ants of South America. Female ant-decapitating flies descend upon a colony, hovering over their targets—which are many times their size—waiting for the right moment to strike. And it happens in an instant: The female fly dives and slams her ovipositor into the membrane between an ant's legs, injecting a tiny egg. (Think of the ovipositor as a needle that instead of delivering drugs delivers a baby. Interestingly, bee stingers are modified ovipositors that instead of delivering babies deliver venom. When a bee stings you, the offender is a female, never a male.)

This does not suit the ants one bit. They sprint around like maniacs, curling up into balls in a pathetic attempt to protect themselves, all the while releasing alarm pheromones, which only makes things worse for themselves. Other ant-decapitating flies are attracted to these scents, and reinforcements zoom in. Soon enough, a swarm of flies is hovering over a battlefield that's littered with wounded ants, still very much alive, yet totally unprepared for what's about to happen to their bodies.

After a few days the egg in each ant hatches into a maggot, which worms its way through the body and into the head. Here it feeds on juices, all the while making sure not to damage its host's brain. The whole time the ant is behaving normally, but in another few weeks it becomes clear why the fly has left the brain alone: This isn't just a host—it's a vehicle, and a vehicle is no good without an engine. The larva releases a chemical concoction that gives it control over the ant's mind, guiding the host away from the colony and down into the leaf litter. It's nice and humid here, the perfect environment for the next stage of the fly's development. (A quick note here on some insect life cycle terminology. The egg hatches into the larva, or maggot, a kind of squishy wormlike thing. It feeds and grows for a while before turning into an inactive pupa, which would be the cocoon stage of something like a butterfly. From the cocoon the insect emerges as an adult.)

Once the ant reaches an ideal location, the maggot releases a chemical that dissolves the membranes holding the ant's various parts together, including the bit that keeps the head on. The noggin pops right off the body with the larva inside, and only then does the fly eat the brain to hollow out the head. When it has had its fill, it clears away the ant's mouthparts, leaving an escape route that it plugs up with a hardened end of its body (remove the pupa at this point and it'll be shaped like the ant's head, a bit like a kernel of corn). The ant-decapitating fly will develop for a few more weeks, all snug in its crib, before emerging from the head and flying away to mate and start the whole weird circus all over again.

All of this makes the ant-decapitating fly a kind of folk hero in the United States. Not because we're impressed by its work in South America, but because we're dependent on its work here at home: Fire ants from down south have invaded the United States and grown enormously successful. Their economic impact each year tallies in the billions, ranging from crop damage to the medical costs for humans unfortunate enough to tango with them (stings result in painful pustules and the occasional severe allergic reaction). The ants seem to have found an unassailable niche in the ecosystem, or so they thought. The government has taken the somewhat desperate step of importing the ant-decapitating fly to wage biological warfare with the fire ant. And it appears to be working, where things like insecticides have largely failed.

Fighting nature with nature is a strange and precarious technique. We've grown so used to dousing our problems in chemicals that it's easy to forget the greatest weapon of all: Every creature has its foe—even apex predators like bears or lions have parasites to worry about. True, introducing an invasive species to combat an invasive species is a risky maneuver, what with them both being *invasive species* and all. But scientists with the USDA determined that the ant-decapitating fly is so specialized, so intent on ruining the fire ant's day and only the fire ant's day, that it was safe to bring to the United States. And with the fire ant laying waste to agriculture and livestock and sending the ecosystem into chaos, not to mention attacking humans, I find it a wee bit hard to feel sorry for the invaders. So welcome to America, little ant-decapitating fly. And good hunting.

Glyptapanteles
Wasp

PROBLEM: I wasn't kidding when I said maggots are helpless.

SOLUTION: A momma *Glyptapanteles* wasp injects her eggs into living caterpillars. After the larvae feed on the juices and burst out of their host, they mind-control the poor beast to defend them as a bodyguard/babysitter.

All of this invading other creatures' bodies and mind-controlling them and dispatching them in horrible ways is relatively rare for a fly. But not the wasps. No, some of nature's most sophisticated parasites are in fact wasps, such as the Ichneumonidae variety, insects capable of brutality like you wouldn't believe.

Indeed, parasitic wasps that inject their young into caterpillars presented Darwin with quite the conundrum: "I cannot persuade myself that a beneficent and omnipotent God would have designedly created the Ichneumonidae with the express intention of their feeding within the living bodies of caterpillars, or that a cat should play with mice," he once wrote in a letter to Asa Gray (whom, if you'll remember, he also complained to about the peacock's tail). Darwin's theory of natural selection posited that a whole lot of creatures must die so others could live, which didn't go over so well at the time, particularly with the religious establishment. These were people with an idyllic conception of nature, who thought that predator and prey could hang tight in Noah's boat and somehow not devour each other before the flood subsided.

The Ichneumonidae wasps have some cousins, though, that aren't about to let caterpillars off so easy: the *Glyptapanteles*. These species don't kill caterpillars outright. Instead, as they erupt out of their host, the larvae bestow the victim with an extra indignity by mind-controlling it into becoming a bodyguard. Thus, under such devoted care, the young wasps are ferried safely into the world.

The female wasp begins by seeking out a caterpillar. And like the ant-decapitating fly, she drills into her victim. Unlike the fly, though, the wasp deposits as many as eighty eggs into the poor critter. These will hatch into larvae and feed on bodily juices, growing fatter and fatter until their host looks like an overfilled water balloon. All the while the caterpillar goes about its life, meandering around and eating—after all, it has eighty-one mouths to feed. But there comes a time in every parasitized caterpillar's life when it must bid farewell to the maggots it has housed, fed, and shared so many memories with. So the larvae release chemicals to paralyze their host and erupt en masse from its body for a full hour, all writhing and wiggling right through the caterpillar's skin. It's not what you'd call a pleasant scene, as almost every square inch of the caterpillar's body is bursting with squirming maggots.

But this isn't the end of the unwitting host. As the larvae grow, each must periodically shed its exoskeleton and grow a new one. This happens several times inside the caterpillar, but most significant is the last molt, which is timed with the larvae's exit (an event known formally and ever so measuredly as an "egress"). As they squirm out, the larvae shed their exoskeletons, which plug up the wounds they leave behind, as it's in their best interest to keep their host alive, at least for the time being. But not all the larvae exit the caterpillar. One or two will stay behind, and these stragglers seem to be responsible for what happens next. By remaining in the caterpillar, those larvae may release chemicals that mess with their host's brain, transforming the caterpillar into an ultraviolent goon that protects the rest of the brood.

So as the caterpillar shakes off its paralysis, it doesn't attack its assailants, and it doesn't flee. It doesn't even have any desire to eat. It stays right where it is as the larvae around it start spinning their cocoons. In fact, it even chips in, spinning a covering over the cocoons with its own silk. And woe to any predators that try to attack the young wasps, including the parasitic wasps that in turn attack these parasitic wasps, because the caterpillar will lash out at anything that gets close with violent swings of its head. Even as swarms of predatory insects descend on what should be an easy meal, the caterpillar stands its ground, swinging to bat away the invaders. One study found that bodyguard caterpillars can fend off 58 percent of predators, whereas nonbrainwashed caterpillars fend off only 15 percent of predators. Not a bad rate of success for a squishy tube of a creature that normally ambles around eating leaves.

Its moment of victory is a short one, though. As the wasps complete their life cycle, emerging from their cocoons and flying off, the caterpillar dies of starvation. It was used, through and through, and it dies an unfortunate death.

I often find myself struggling to find words to describe how amazing this sort of thing is, but since I'm a writer and all, I should probably give it a shot. The majesty of a one-hundred-foot blue whale is one thing, and the beauty of a brilliantly iridescent blue morpho butterfly is another, but how astounding is it that a creature has evolved to mind-control another into doing its bidding? And ant-decapitating flies and *Glyptapanteles* wasps aren't anywhere near the only animals to do this—all manner of critters have independently arrived at the mind-control solution to their problems.

But such complexity is, weirdly enough, the product of the simple process that is natural selection. Wasps didn't suddenly arrive at the idea of assuming control over caterpillars. The same way the mustache toad developed its weaponized mustache, over generation after generation, wasps with certain variations were more successful at reproducing. They may have begun, for instance, by injecting caterpillars in the first place and hollowing them out. That certainly helped. (Indeed, there are any number of parasitic wasps that stop there, having for whatever reason not evolved further complexity.) Then other variations resulted in the larvae avoiding major organs, guaranteeing a fresh meal for longer. That, too, helped the young better survive, and therefore pass down the genes responsible for such variations. Small modifications accumulate over time to add up to an awesome event like the mind controlling of a caterpillar.

Evolution, though, is a two-way street. Creatures can evolve this kind of onslaught, but their victims can evolve a defense. And there's one caterpillar out there that isn't going to sit by and suffer the indignity of predation. No, that won't do.

IF THE BODY OF ANOTHER CREATURE FITS, WEAR IT

Different species arriving at similar adaptations like mind control is known as convergent evolution. A good example is flight. Birds evolved from dinosaurs into critters capable of flying, but bats—which are mammals—also evolved this ability, only they use a membrane of skin as a wing instead of feathers. These are different lineages, but both bats and birds found it advantageous to take to the air. The same goes for the many creatures that have stumbled upon mind control. As the saying goes: If the body of another creature fits, wear it.

Asp Caterpillar

PROBLEM: Slow, plump caterpillars are easy meals.

SOLUTION: The asp caterpillar grows a groovy, irritating hair-do with secret weapons underneath: spines that sting with such ferocity that they can trigger breathing difficulties (and panic attacks) in humans.

Having already dumped her husband some years earlier, naturalist Maria Sibylla Merian at last gave in to an overriding obsession with insects, taking her youngest daughter and traveling from Amsterdam across the Pacific Ocean to Suriname. Here she spent two years traipsing through jungles collecting bugs, taking them home, and raising them—studying them, drawing them.

Few caterpillars she found were as bizarre as the flannel moth, with its shock of hair (an entomologist friend of mine rightly points out its resemblance to Donald Trump). "The skin of these caterpillars, under the hair, looks very like human skin," Merian wrote. "They are very poisonous; if one touches them with the hand, it swells up immediately and is very painful, as I discovered myself." And poor Merian didn't have the luxury of modern medical treatment. You see, the year was 1699, more than 130 years before Darwin stepped foot in the Galápagos. Merian was literally a trailblazer in a man's world, a woman whom science has largely forgotten. And she was probably the first natural historian, regardless of sex, to undertake a scientific expedition to a foreign country.

Merian was in Suriname to expose the mysteries of insect metamorphosis, which in the seventeenth century was almost totally unstudied by modern European science, itself in its infancy. The big questions: Why should butterflies and moths bother with such extreme transformations? Why not hatch into a butterfly and forgo all the caterpillar business?

Well, it turns out it has a lot to do with food. By taking on such a radically different form, the caterpillar can exploit a niche different from the adult's, so it doesn't need to compete. But this life cycle leaves the caterpillar—a piece of meat just asking to get assaulted, as the *Glyptapanteles* wasps have noted—vulnerable. Unless, that is, it evolves a defense.

Among the many species whose caterpillars have adopted weaponized hairdos, none is more powerful than the one Merian tangled with: the asp caterpillar, aka the puss caterpillar, of the Americas. Yeah, there's a bit of a disconnect between the names "asp caterpillar" and "puss caterpillar." But then again, there's a bit of a disconnect between how the caterpillar looks and what it can do to the human body. Reports of envenomations vary, likely due to the dose of venom received, but a sampling of reactions include: "It immediately felt like a hammer hit me" and "it felt as though my arm had been broken" and "I have had kidney stones before, but I believe the pain I am experiencing from the asp sting is worse." Symptoms can range from faintness to a burning sensation, fever, and, rather vaguely and intriguingly, "abdominal distress." The paper that reported these cases noted that victims, not suspecting that a caterpillar could cause such suffering, are often "launched into panic attacks by the unexpected onset of the pain."

Really, the asp caterpillar is a crawling head of hair—flowing, golden locks that make the creature look like it fell asleep under a blow-dryer. Like other caterpillars, it goes through molts as it develops, in which it sheds its exoskeleton and grows a bit bigger. But at each stage, the asp caterpillar adopts an ever more luscious hairdo, until it ends up with a wispy, inimitable 'fro. Yet it's not the hair that'll ruin your day. It can irritate your skin, sure, but it ain't got nothing on what lies underneath: spine after spine after spine, each attached to a venom gland, and each capable of delivering toxins to those unfortunate enough to assault the creature.

And asp caterpillars can appear in expansive numbers. In Texas they're a particular problem, where the things have been known to blanket trees. Officials have had to shut schools down because children, thinking the fuzzy caterpillars must be cuddly, pick them up. In 1923 the USDA noted: "In Dallas and other Texas cities hundreds and even thousands of cases of stings have occurred during a single season, and in some cases the fear of the caterpillars became almost a mania owing to the description of the effect of the stings which was passed from one to another." The report's author, speaking like someone who had never been the victim of a caterpillar, added: "It is possible that the effects of stings may be made more serious by the hysteria engendered by these often exaggerated statements, especially if published in newspapers."

One of the more elaborate accounts of an asp caterpillar sting comes from North Carolina in 1997. A man was cleaning fish when

**YOU KNOW,
NOW THAT YOU MENTION IT . . .**

The asp caterpillar sports what are known as urticating hairs, meaning they're irritating to the skin and mucous membranes. Tarantulas also have them, kicking the hairs off their bums when threatened, to create a cloud of misery. A twenty-nine-year-old British man found this out the hard way when in 2009 he visited the ophthalmologist complaining of an itchy, watery eye. Only when the doctor informed him that he had tiny hairs embedded in his eyeball did the thought occur to him that maybe, just maybe, it was the dose of urticating hairs his pet tarantula had hit him with *three weeks earlier*. He'd been cleaning its tank with the tarantula still in it, when the predictable ensued. As the doctor recounted: "He turned his head and found that the tarantula, which was in close proximity, had released 'a mist of hairs' which hit his eyes and face." Having treated the patient, the doctor seemed happy to add that the man "now wears eye protection before handling the tarantula."

he reached into his cooler and felt a severe burning in his arm. Sitting there on his limb without the slightest care in the world was an asp caterpillar. With admirable foresight, the man showed up to the hospital carrying the caterpillar "in a cellophane-covered plastic cup." It was a good effort, though it wasn't as if the doctors could do anything special if they knew what attacked him—the caterpillar's venom isn't nearly as well understood as that of something like a rattlesnake because, let's face it, there aren't many grad students lining up to devote their lives to studying hairy caterpillars.

Upon arrival, the man told the staff he was experiencing the worst pain he'd ever felt, noting with what I imagine was total sincerity: "It feels like my chest is caving in." What follows in the account is mostly a hodgepodge of medical terminology, but here are a few highlights. Initial symptoms were dry mouth (eh, no big deal), vertigo (okay, getting serious), and difficulty breathing and swallowing (definitely serious). Doctors pumped the man full of pain meds, plus some antispasmodics to treat the twitching in his upper leg, and then some more pain meds to be sure. Thus, filled to the brim with drugs, our brave patient stabilized, no doubt with a finer appreciation of caterpillars.

Other critters have not only taken notice that it's not a solid idea to mess with these things, but that stinging caterpillars got it made. One bird, the cinereous mourner, even seems to mimic hairy caterpillars. As a chick, it develops long orange plumage that bears a striking resemblance to the outfit of a stinging caterpillar it shares a habit with. Nature, it seems, is full of creatures more than willing to bite your style.

Ocean Sunfish

PROBLEM: With so many predators about, releasing eggs into the open ocean is asking for trouble.

SOLUTION: If you're the ocean sunfish, you subscribe to the spray-and-pray method of reproduction, dropping a world record 300 million eggs in a single go. After all, someone is bound to survive.

It's a brisk-but-not-uncomfortably-brisk January morning, and marine biologist Tierney Thys is on a lawn kneeling over the biggest bony fish in the world. Well, this particular ocean sunfish could well have been, had it not died just a year into its life and washed ashore. The animal is flat and circular, a bit bigger than a dinner plate. Its eyes are gone, probably stolen by gulls that couldn't peck through its thick skin to reach any other meat. Also missing are its two towering fins, which normally sprout from its back and belly. Those were probably snagged by sea lions, which have been known to nip them off and then toss the fish back and forth like a Frisbee.

Gathered around Tierney are two dozen teachers, who are ostensibly here to learn how to better communicate science to their kids, but who now appear to have *become* the kids, giggling and oohing and aahing and snapping photos. Tierney plunges her scalpel into the sunfish, cuts away a chunk of the sandpapery skin, and passes it around. One teacher asks her to walk us through the digestive system so she could film it and show it to her kids, whom she's teaching about such things at the moment. Tierney obliges, pointing out its stomach and lifting up its pearl white intestines.

But I'm more interested in the genitals. I can't say that in front of two dozen teachers, so I wait for them all to file back inside for their conference, and then ask Tierney if it's a boy or a girl. I'm disappointed to learn that it died too young to tell such things, for I wanted it to be a girl, and I wanted to see her eggs. The ocean sunfish, which grows to ten feet long and two and a half tons, can release 300 million of the things, more than any other vertebrate. And that was estimated from the ovaries of a four-foot female, which hadn't even grown to half her potential size.

The ocean sunfish takes a laid-back approach to whoopee: spray-and-pray reproduction. Because the female is so enormous, and her eggs are so tiny—about the size of a BB gun pellet—the sunfish can pack a ton of them into her body. And it's a good thing she can. Like most fish, the sunfish utilizes external fertilization, with the female dumping eggs into the water as the male dumps his sperm. With any luck, the two parties will meet and fertilization will commence. Afterward, the female doesn't care for her kin in the least bit. She disperses her millions upon millions of eggs and swims away.

Thus the ocean sunfish plays no small part in contributing to the biomass known as plankton. Plankton are teeny-tiny lifeforms at the mercy of currents. Think microscopic plantlike organisms, called phytoplankton—whose photosynthesis is responsible for producing half of the oxygen in our atmosphere—as well as animals, called zooplankton. Then there are the eggs like the ocean sunfish's. Collectively they are a monumentally important source of food in our seas for all manner of predators, from tiny shrimp on up to the sixty-five-foot-long whale shark (which, as it happens, is the biggest fish, bony or otherwise—it has a skeleton made of cartilage instead of bone). And that's a problem if you're a sunfish egg that'd prefer surviving. There's so many of them, though, that a few are bound to make it.

THE UNGLAMOROUS LIFE OF THE PENIS-FACE FISH

There are a few weirdos when it comes to fish sex, such as the penis-face fish of Vietnam. The male has an organ on the underside of his face, with a serrated, hooked bit that grasps on to the female as another bit shoves a sperm bundle into her naughty parts, which are also on her throat. Oh, I should mention. They defecate out of their face sex organs as well. So now that's something you know.

If the eggs can manage to develop into larvae they have another trick to make it into adulthood, one that gives away their heritage. Ocean sunfish, as it would turn out, are highly modified puffer fish, which long ago left the reef for life in the open ocean. Adult ocean sunfish may have lost their spikes, but the larvae have retained the weapon of their ancestors. The youngsters are almost spherical, with doe eyes and huge conical spikes jutting in all directions. It isn't hard to see just how beneficial these spines would be out in the vast sea, for there are no reefs to cower in or sand to burrow into. There's only hungry mouth after hungry mouth, which may not appreciate spiky prey. As the sunfish develops,

INTO THE BELLY OF THE BEAST

When he was sailing around the world on the *Beagle*, Darwin, off the coast of South America, was "amused by watching the habits" of the porcupine fish, a species related to the puffer fish, with the same ability to inflate itself, thus deploying those famous spines. He was under the mistaken impression that should a shark snag one of these critters, it wouldn't be the spikes the predator would have to worry about as much as the porcupine fish gnawing its way "through the sides of the monster, which has thus been killed." That's a nice underdog story, but it isn't so much possible. Darwin would return to England and come up with the best idea of all time, so we'll go ahead and give him a pass on this one.

these spikes shape-shift to become thinner and pointier, until they disappear almost entirely as the larva enters the adult form. The minute projections remain, though, helping give the sunfish's skin that sandpapery texture (it's still rough enough to rip up human flesh, as Tierney, who spends her days tagging the sometimes uncooperative creatures, can attest).

When the sunfish tops out at ten feet long and thousands upon thousands of pounds, it'll set yet another record in the animal kingdom. From a larva measuring a fraction of an inch long, it will have grown in size 60 million times, the most epic growth among the vertebrates. That's like you coming out as an eight-pound baby and developing into a human that weighs a half billion pounds.

Yet of the 300 million eggs the female sunfish releases, only a fraction will survive long enough to realize that growth. We know this because, for one, if a lot made it, the oceans would be lousy with sunfishes. And also because although populations of a given species may fluctuate—say, due to the arrival of some cataclysmic outside force or the departure of a competing species—on average only two of an animal couple's offspring will survive to reproduce, thus replacing their parents in the population (we humans and our explosive numbers are a notable exception, having excused ourselves from the food chain). Just two. In the case of the ocean sunfish, two of 300 million potential fish.

But in the end, the mama's big bet pays off. She doesn't stick around to watch her kids dodge their first whale shark attack or lose their first spikes or finally become adults with responsibilities and stuff, but in her own way she ensures their survival, however few of them make it. Her solution to the problem of predation is to overwhelm the ecosystem with sheer numbers. It ends up being a massacre, sure, but it's a massacre with a happy ending, which is the best kind of massacre of all.

Lowland Streaked Tenrec

PROBLEM: The forests of Madagascar are home to plenty of hungry mouths.

SOLUTION: The lowland streaked tenrec is the only mammal that can communicate with stridulation, rubbing quills on its back to make chirping noises that reel its straying kids back in.

There's no better way for a parent to tell his or her child "I don't trust you and I trust myself even less" than putting the child on a leash. I mean, I kind of get it. You wouldn't want to lose your kid. We are, after all, mammals that don't have the luxury of pumping a few hundred million eggs into the world and hoping for the best. Human moms don't have that many chances to pass their genes along—but let's be real, maybe a leash isn't the most elegant way to hold on to your kids.

In the absence of leashes and the opposable thumbs required to operate them, an enchanting mammalian mother in the forests of Madagascar, called the lowland streaked tenrec, has hit upon a more punk-rock solution to losing her meandering kids as the family forages at night: spikes. While the tenrec is streaked with bright yellow and black spines, making it look like a hedgehog that dressed up as a bumblebee for Halloween, there's a cluster of thirteen to sixteen specialized spikes running down the mom's back that she rubs together to produce a high-pitched noise, which sounds a bit like rapidly running your finger up and down the teeth of a comb. It's the same principle behind a male cricket chirping by rubbing his wings together in order to attract females, but no mammal other than the tenrec can communicate in this way, a technique known as stridulation. But then again, there's no mammal quite like the lowland streaked tenrec.

Tenrecs have done well for themselves on Madagascar. The island was once part of the supercontinent Gondwana, but set out

SO HOW WILL WE KNOW WHEN THERE'S AN AWKWARD SILENCE?

There's a tradeoff to all this commotion: Stridulation helps the tenrec find its family and the cricket find a mate, but it can also attract unwanted attention. A remarkable illustration of this comes from the field crickets of Kauai. In the 1990s an invasive parasitic fly arrived on the island, a menace that homes in on crickets by their chirping and deposits its kids on their backs—maggots that then burrow into their hosts and eat them from the inside out. At some point, though, a mutation arose that made male field crickets incapable of chirping, rendering them largely invisible to the fly. Natural selection therefore favored those crickets, as their chirping colleagues perished: In only twenty generations the population had almost entirely shifted to nonchirping males. However, this in turn made them less conspicuous to females, which love themselves a good chirp. It seems the females arrived at a compromise, accepting the mutes as they are, probably because they have no choice on account of all the chirpers being eaten alive.

Who says romance is dead?

■ THE WASP THAT BRAINWASHED THE CATERPILLAR

on its own almost 90 million years ago. Some 30 million years later, a small mammal made it to the island, splitting into different species of tenrec that assumed different niches. There's a water-loving tenrec with webbed feet, for example, while other tenrec species take to the trees. The lowland streaked tenrecs became nocturnal, social foragers, forming into groups of as many as two dozen individuals to root around the forest floor with their long schnozes, all the while communicating in the darkness by stridulating. A momma lowland streaked tenrec can have eleven kids, quite the handful when you consider that tenrec clans fan out as they forage, and each infant may wander ten feet away from the mother. So amid the racket of the rain forest, the distinctive note of stridulation acts as a beacon for any lost children.

The scientist who revealed much of this strange communication, in the 1960s and '70s, was a man named Edwin Gould. He had a hunch that stridulation was a mother's way of keeping track of her kids, so he set up a square arena with speakers in two corners, dropped a young tenrec in, and put on the sweet sounds of rasping quills. Of the nineteen infants he tested, fifteen made a beeline for whatever speaker was active. In a second experiment, this time in an outdoor enclosure, he immobilized a mother tenrec's stridulating spines by slathering them in glue. While youngsters typically feel free to wander, now they followed much closer, likely using scent instead of the stridulation to pinpoint her position.

But this is more than a matter of not getting lost, for the tenrec isn't the only predatory mammal that's made it to Madagascar. There's also the Malagasy ring-tailed mongoose, with its gorgeous alternating bands of black and burnt orange to rival the lowland streaked tenrec's own lovely color scheme. Also stalking these forests is the fossa (pronounced FOO-suh), which looks like a cat but is more closely related to mongooses. Present the odor of either of these predators to a tenrec, as Gould did, and you'll get yourself quite a reaction. The nonstridulating spines at the back of the tenrec's skull, which normally lie flat, perk up and over the head into what looks like a spiky, bright yellow lion's mane, as the critter emits a *putt-putt* sound. If whatever is threatening it isn't impressed, the tenrec escalates the noise to a kind of crunching and starts bucking its head, driving the detachable quills into the villain's snout or paw.

Yet the lowland streaked tenrec's sonic modus operandi gets even weirder: Gould ran more experiments that showed the critters could be using echolocation to navigate. In a pitch-black room he placed a disk on a pole four and a half feet off the ground, and just below that another platform with a ramp that connected to a box containing a reward of food and water. He then assembled two groups of tenrecs, one that he outfitted with tiny earplugs and one without. Their goal was to drop down to the secondary platform and scurry to the reward box. The test subjects without earplugs did great. The others . . . not so much. Gould found that tenrecs with earplugs came to the edge of the disk to search for the platform far more often and spent far more time doing so, often missing the mark entirely when they made the leap.

Yeah, they could have in part been sniffing out the location of the reward box, but this experiment showed that sound clearly plays a role in their navigation—and Gould got it all on a tape recorder. It turned out that the lowland streaked tenrecs weren't stridulating to find their way around, but clicking their tongues, often in rapid succession. And not just on that platform: Gould noted that one enterprising little tenrec escaped in his laboratory, tongue-clicking as it scurried about. Thus, like bats, tenrecs appear to echolocate, sending out sound waves and listening as they bounce back to help build an image of the environment.

And so our little tenrec mother, an unlikely master of sound, wanders the dark, perilous forests of Madagascar, clicking and stridulating her heart out as her kids amble nearby. But on the other side of the world in South America, there's a toad that would be appalled at such irresponsible parenting. Because she prefers to hold her kids tight.

Really tight.

Surinam Toad

PROBLEM: A pond, with its attendant predators, is a dangerous place for a tadpole.

SOLUTION: Instead of leaving her kids at the mercy of the fiends, the Surinam toad mother embeds her eggs in her back and lets them develop under her skin, before the young erupt in unsettling fashion.

The central irony in my life is being aware that as a biological entity, my sole purpose is to reproduce and pass along my genes, yet at the moment I haven't the slightest desire to do so. And I'm in my early thirties. My younger sister has an adorable toddler, and I'm pretty cool with him. He even smiles at me sometimes. But what tends to keep me from having kids of my own is stuff like this: My sister once sent me a picture of my nephew in the backyard holding on to the lawn mower (he's fond of such machines), butt-naked and with a turd behind him, looking over his shoulder at the camera with a knowing look on his oversized toddler noggin of: "Yeah, I just did that." It seems that right before bath time, he'd caught sight of the apparatus out back and made a run for it. Holding on to the lawn mower, with the wind hitting his bare ass, he was apparently so excited that he lost his self-control. Which is all to say that perhaps picking up a kid's doo-doo off the lawn isn't on my short list of things to do in life.

What a trifle that is, though, compared to the trials that the Surinam toad mom must suffer. She doesn't lay eggs and leave them for predators to pick off. No, this toad will not part with her young

■

so easily. After an underwater mating session of somersaults and egg laying with a male strapped onto her, the female's fertilized eggs end up on her back, where skin grows around them, providing a cozy sanctuary for them to develop in before they erupt out of her en masse in what can only be described as a not okay thing for human beings to witness. (In her South American travels, Maria Sibylla Merian came across such an event. In one of her paintings from the expedition, a momma toad loaded with kids swims in a pond as a youngster that's broken free trails behind. Next to them is a lovely plant and two shells, as if Merian is trying—and failing—to lighten the mood.)

The husband and wife team of George and Mary Rabb published what is probably the most superb account of the somersaulting sexy time of the Surinam toad—if you can get past sentences such as "The ascent in the turnover was rotation about the longitudinal axis and in the descent they rotated transversely." It all began on a spring night at the start of the swinging sixties in the Chicago Zoological Park. The lovers were two unnamed Suri-

TOADAL RECALL

What's the difference between a frog and toad? you might ask. To which I'd answer that I'm not sure. No one is quite sure. Toads tend to have warty, drier skin and shorter legs, but there are all kinds of exceptions. Really, it's a problem of semantics: Animals didn't evolve for us to place in categories. Toad or frog, frog or toad, it makes no difference to Mother Nature. The Surinam toad is so named because it lives in Suriname and it's pretty toadish. And that, quite frankly, will have to do.

nam toads, who wished to remain anonymous, or whose names have simply been lost to history, or who perhaps were never given names in the first place because they were toads.

The coupling began with a kind of mounting typical for frogs and toads, known as amplexus, with the male holding tight to the female's lower abdomen with his forearms. For hours the pair did nothing but periodically swim up for air, at times rearing back into somersaults before they could break the surface. There were no sperm or eggs during all of that, but the researchers noted a "noticeable tumescent build-up of the female's back skin." She'd grown ready to accept the eggs, and finally, after the toads completed nine unproductive somersaults, George and Mary glimpsed the first egg stuck to the female's back. The toads continued their flips, with the female laying three to five eggs at each somersault apex. At this point the eggs dropped onto skin folds on the male's stomach, and he released his sperm to fertilize them. As the pair descended to right themselves, the eggs would drop onto the female's back, where the male helped implant them with a bump of his belly. They somersaulted again and again, and at the end of their epic twenty-four-hour mating session, fifty-five eggs were embedded in her back, while only eleven had missed their mark and sunk to the bottom of the tank.

When it was all over, the Rabbs removed the male and dropped him into another tank with more females. He tried his luck with them as well, but they rebuffed him. "While he was attempting to clasp the other females," the Rabbs write, "the skin of his hind legs became loose and hung about them like ill-fitting stockings. The shedding proceeded over the entire body, and he ate the skin as fast as he could get it off in a grotesque sort of ballet." They fail to mention, though, the grotesqueness that the female was about to endure.

The swollen skin on the mother's back starts to grow around the eggs, not entirely enveloping them, but embedding them snugly in a honeycomb pattern. They'll develop there for several months, until they initiate one of nature's more disturbing hap-

THE PITFALLS OF SPECIES AND AUTOCORRECT

Since I now can't stop thinking about the limits of human language, this is as good a time as any to talk about the notion of a species. Agreement among biologists as to what exactly a species is can be such a problem that it's known as the species . . . problem. Oxford Dictionaries defines a species as "a group of living organisms consisting of similar individuals capable of exchanging genes or interbreeding." Yet there are all manner of hybrid species out there: a grizzly and polar bear, for instance, can come together in the wild to produce offspring. They're different "species," but are capable of exchanging genes, which goes to show that we can do our best to make sense of the natural world, but in the end it's an issue of language.

Further semantic conundrums arise when scientists try to name "species" that are the product of "species" mating with other "species." That grizzly/polar bear baby can be called a grolar bear or a pizzly, both of which I object to because my word processor tried to autocorrect them and therefore screwed up my writing flow, which I pray you didn't notice.

penings. One by one the young toads—up to a hundred of them if the mating was good—begin to emerge. They haven't grown into tadpoles at all, a rarity among frogs and toads known as direct development, but instead miniature versions of their mother, poking their skinny little fingers out of the holes in her back. As they squirm under her skin, sometimes all at once in a wave of twitching, every so often one pops out and swims away. And so they escape, tiny frog by tiny frog, until the mom is left with an empty nest and pockmarked skin that she eventually sheds. It's a radical solution to the problem of predation, to say the least. But if any villains want to get at those kids, they've got to go through Mom first. Like, actually through her.

When the little Surinam toads head out into the big bad world, a world swarming with all kinds of dangers, it helps that they're bizarrely flat to blend into the leaves at the bottom of their pond, spending their time splayed out among the sunken vegetation. Still, they're vulnerable to predation and will be lucky to one day do sexual somersaults.

But what if they had some kind of shelter? What if they could build a home? What if they made a home out of a sea cucumber's anus?

CHAPTER 3

You Need a Place to Crash

*In Which Fish Swim Up
Sea Cucumbers' Bums and
Birds Build Nests So Big
They Pull Down Trees*

Shelter isn't just a human pursuit. It can be nasty out there, with lots of teeth and claws and such things flying around. So certain enterprising creatures have hit upon solutions to the housing crisis. Birds build their nests, and ants dig their tunnels, but some animals opt to make their homes out of . . . other creatures. Hey, at least it's always warm.

Pearlfish

PROBLEM: Fish don't have many places to hide from their enemies on a barren seafloor.

SOLUTION: The pearlfish swims up a sea cucumber's anus, makes itself comfortable, and eats its host's gonads.

They say home is where the heart is, but for the pearlfish, home is more like where the gonad is. Because to find shelter it wiggles up a sea cucumber's anus. And lives there. And eats its gonads.

It all begins innocently enough. The skinny, eel-like fish approaches the sea cucumber and gives it a sniff, moving up and down the length of its soon-to-be victim with its body pointed almost vertically. It's looking for the sea cucumber's breath, because, well, sea cucumbers breathe through their derrières. If the sea cucumber detects the pearlfish, it'll hold its breath, sealing its butt like a human holding in a fart. But the sad sack is only delaying the inevitable. At some point it has to breathe, and that's when the pearlfish strikes.

Once the pearlfish sees its window of opportunity, it has to make one of the toughest decisions in the animal kingdom: whether to enter the sea cucumber's bum tail first or head first. This, of course, all depends on the size of the orifice (called a cloaca, since it's used for not just waste removal but also reproduction and breathing). If the opening is big enough to enter head first, the pearlfish goes at it full tilt, jamming its face in and rapidly flicking its tail to fire itself into the sea cucumber. If the opening is too small, the fish first inserts its thin tail, then backs in slowly. And because they're so much smaller, juvenile pearlfish will go in head first 80 percent of the time, whereas adults go in *tail first* 80 percent of the time (yes, someone calculated that).

You wouldn't blame the pearlfish for not wanting to hang out in its host's intestines. So instead, once the fish is in there it wiggles into what's known as the respiratory tree, which branches out as tubes on either side of the sea cucumber's intestines. The sea cucumber breathes by pumping water in and out of its cloaca and into the tree. So not only is the pearlfish enjoying shelter while it's in there, it's getting a constantly replenished source of water, and therefore oxygen.

Some species of pearlfish are content sitting there biding their time, safe from predators circling above. Others . . . they get a bit hungry. They'll start feeding on various parts of the sea cucumber, including the gonads, which isn't as bad as it sounds. Well, I don't

WHY YOU'LL NEVER MEET A SEA CUCUMBER DENTIST

Sea cucumbers aren't entirely defenseless against the onslaught of the pearlfish. Certain species are equipped with calcified projections known as anal teeth, which are shaped like cones. They all point inward toward the center of the anus, looking a bit like the doors they have in sci-fi movies that don't swing or slide open but open up from the center out, as a camera shutter would. These may well have evolved specifically to help keep pearlfish out, which goes to show that at least a few kinds of sea cucumbers find it less embarrassing to have teeth in their bums than having their gonads eaten.

think the sea cucumber particularly enjoys the gesture, but once the gonads are gone it can grow new ones. The same goes for its other organs: The critter is a remarkable regenerator. But it can find itself overwhelmed with enemies, and while it can regrow the bits one pearlfish has eaten, another is bound to show up and hack away at its insides once again. In fact, a biologist once found fifteen pearlfish in one poor bastard of a sea cucumber.

And if you thought the fish were ignoring each other's advances in there, you've got another thing coming. Pearlfish also get busy in the respiratory trees, the ultimate indignity after the sea cucumber has itself been sterilized. This sort of thing is not uncommon: A lot of parasites sterilize their hosts, and it makes good sense. It takes a tremendous amount of energy and resources to reproduce—remember the lengths the flatworm goes through to avoid bearing young. If a parasite can figure out how to sterilize its victim, that means more energy is available to support its own shenanigans. Whether sterilization is the pearlfish's strategy or the gonad happens to be tasty isn't yet clear. Could be both, really. You'd have to ask a pearlfish.

Now, while we use shelter more to protect our relatively hairless and fragile bodies from the elements, other creatures, including any number of burrowers, use shelter to avoid predation. The pearlfish is doing the same thing, really, only it's burrowing up sea cucumber cabooses. You have to consider that the seafloor is a very dangerous place indeed. It's not all intricate coral reefs out there, with plenty of hiding places like caves and crevices. Much of the ocean is featureless stretches of sand.

And pearlfish don't limit themselves to taking up residence in sea cucumbers. Some species instead invade sea stars, and still others go after bivalves like oysters. Those strategies would appear to be better options, considering the sea cucumber itself is vulnerable, on account of being essentially a tube of meat wandering along the seafloor. But while it may look appetizing, in reality, it's not, for the sea cucumber has a secret weapon: It disembowels itself to scare away predators.

SUFFERING FOR FASHION

Pearlfish are called pearlfish because folks will pop open oysters and find the fish covered in the beautiful material that the mollusk would typically use to form a pearl—a nice fashion accessory if the fish weren't too dead to enjoy it. (Any pearl, by the way, is a shelled mollusk's defense against a foreign object that has infiltrated it. By coating the invader, the mollusk isolates and neutralizes it.) Researchers have found that when these fish are actually alive in there, they're rapidly vibrating their swim bladders, like the toadfish, to call to their compatriots swimming out in the open. It might seem like the inside of an oyster wouldn't be great for acoustics, but in fact the shell amplifies the frequencies the fish are calling at. So it's like a microphone. That you live in. And probably isn't too happy with you.

Admittedly that doesn't sound like an effective defense at face value, but hear me out. When a sea cucumber feels threatened, the connective tissues that hold its organs in place soften in an instant. Depending on the species, the body wall around either the mouth or the heinie also softens. With everything loosey-goosey, the sea cucumber contracts its muscles, firing its guts through the weakened body wall at one end or the other. And it isn't just the intestines that take these explosive little field trips. Sometimes even parts of the respiratory tree and, yes, the gonads can go, too (it seems that there are few jobs in the animal kingdom as thankless as that of the sea cucumber gonad). There are some species that have evolved an even more potent defense, ejecting toxic tubes out of their anuses along with the guts. The explosion is a kind of surprise that maybe, just maybe, is weird and smelly and goopy enough to scare away predators. And as traumatic as it all sounds, sea cucumbers can regenerate the lost organs in a few weeks.

Interestingly, the pearlfish doesn't seem to trigger these defenses, and as yet no one is quite sure why. Maybe it's more energetically cost-effective for the sea cucumber to deal with regrowing gonads than the intestines and all the other ejected bits. So what the pearlfish ends up with is a living, breathing mobile home, which staves off attack by dropping its guts out, like a James Bond car deploying caltrops to slow pursuers. Except the sea cucumber isn't going to get the girl. You know, on account of missing its gonads.

Tongue-Eating Isopod

PROBLEM: Floating around in the open sea is perilous (just ask the ocean sunfish).

SOLUTION: A certain crustacean swims into a fish's mouth and doesn't stop there—the tongue-eating isopod didn't get its name for nothing.

Living in a sea cucumber and eating its gonads is nice and all, but what if you're looking for a place with more of a view? Perhaps where you can experience the big blue ocean in all its glory? You'd think the luxury would come at a price. But no, on the contrary. If you can manage to find a place like this, you're going to get away with murder.

Such is the existence of the tongue-eating isopod, *Cymothoa exigua*, an ill-behaved crustacean that invades a fish's gills and latches on to its tongue, consuming it and replacing the organ with itself. Here it'll work as a prosthetic tongue, staring out of the fish's mouth with one hell of a view (well, until it goes blind, at least—more on that in a second) before making its exit and leaving the fish to starve to death. Having trashed its home, the isopod may lose its security deposit, sure, but it gains the ability to pass its genes down to the next generation.

First, the isopod must find its victim in the vastness of the sea. Juveniles only have so much energy to burn, so while they spend the first few days of their lives swimming around searching frantically for a fish to invade, if they're unsuccessful they'll switch to passive mode and wait for a host to approach, then ambush it. They're looking for chemical cues, and when the passive juveniles catch a whiff of something, they'll whirl back into action and look for silhouettes passing against the sun above. Target acquired, the isopod makes a break for the fish and wiggles its way into the gills.

All tongue-eating isopods are born male. When one arrives in a fish and finds that none of its compatriots are there yet, it stays male. But if another male arrives, the first isopod switches into a female and makes its way onto the tongue, digging into the tissue with seven pairs of incredibly sharp claws. She'll live here for the rest of her life, so she has no use for eyes, which fade away as she develops. She also loses the ability to swim.

She begins feeding with five sets of jaws, a few of which are shaped like needles, slicing open the tongue and sucking out the blood. Slowly the tongue atrophies away and the isopod takes its

A VALENTINE'S DAY MOST TRAGIC

Isopods get *way* bigger than the tongue-eating variety. The giant isopod sticks to the deep seafloor, dining on carcasses falling from above and growing to over a foot and a half long. You never know when you're going to stumble across your next meal down there, and accordingly the giant isopod can go a very, very long time without eating. One individual in captivity in Japan, known somewhat half-assedly as "Giant Isopod No. 1," went on a five-year hunger strike before dying on Valentine's Day, 2014. Perhaps it was lovesick, or maybe it had something to do with not eating for five years.

place, as the fish uses the parasite to grind food against the roof of its mouth. Thus the isopod, like the caterpillar-enslaving wasp, is able to keep its host alive and well to support its own reproduction, for the time being, at least. Plus, the parasite gets both food and shelter.

But what's the point of the sex changing? This happens to be a rather brilliant strategy in the open ocean for a creature that's dependent on another to survive. If two isopods are lucky enough to wind up in the same fish, it's all for naught if they're both female or both male. And so the tongue-eating isopod has evolved to break the rules, switching sexes to help guarantee the ability to mate.

When the blind female is ready to release her young, she may be able to somehow detect when the fish is schooling and do it then, saving her kids the grief of wandering around searching for a host. Now finding herself in an empty nest, the new mother finally releases her grip, either popping out of the fish's mouth and sinking like a stone or allowing the fish to swallow her. Not that the meal is going to do the fish any good. Unable to feed without a tongue, it'll starve to death.

Weirder still is the lifestyle of one of *Cymothoa exigua*'s cousins, *Cymothoa excisa*. Unlike the former, the latter doesn't totally destroy its host's tongue and replace it. (And even then, *Cymothoa exigua* doesn't necessarily always do that. It targets several kinds of fish, but only fully replaces the tongue of the rose snapper. Why that is isn't yet clear.) Instead, *Cymothoa excisa* politely sips on its host's tongue blood, allowing it to live. This means it gets extra time for mating, so more and more males show up in the gills. Indeed, they may be *getting in line*. When the female eventually dies, the male that mated with her may then turn into a female and move onto the tongue, and the next male in line steps up to mate with her, creating a perpetual chain of misery for the fish.

If that all wasn't bad enough, it seems that overfishing is making matters even worse for fish species that fall prey to tongue-eating isopods. A 2012 study found that a population of fish in a protected environment in the Mediterranean were infested 30 percent of the time, compared to almost 50 percent of their counterparts in heavily fished waters. It may be that in these times of overfishing, natural selection has been favoring smaller fish that reproduce faster, which have an edge on their slower-growing

A KAMIKAZE CRICKET MOST INFESTED

A parasite like the tongue-eating isopod that eventually kills its host is known as a parasitoid. Keeping its host alive can be in a parasite's best interest, on account of it providing food and shelter, but once the invader is able to reproduce, fulfilling its purpose in life, the host will have fulfilled its own purpose and the parasite is free to dispatch it (and technically become a parasitoid).

Still, though, some hosts can end up surviving horrific trauma. There's a parasite called the horsehair worm, for instance, that invades crickets and mind-controls them to kamikaze into water. At this point the foot-long worm erupts out of the cricket's exoskeleton and squirms away. In one lab, a scientist witnessed thirty-two such worms piling out of one sad, sad cricket, which somehow survived. Intriguingly, there's a six-foot species of horsehair worm out there, but no one knows what kind of giant insect it parasitizes. But boy, that giant insect certainly knows it.

comrades. And it could be that these smaller fish are unable to mount good defenses against the isopods, ending up infested more often.

So envy not the fishes. Although they freely swim Earth's beautiful oceans, they also have to deal with not only us, but one hell of a parasite. Feel free to envy the tongue-eating isopod, though. Free room and board with an incredible view? We could all only be so lucky.

Pistol Shrimp

PROBLEM: The seafloor is a war zone.

SOLUTION: The tiny wonder known as the pistol shrimp forms huge societies ruled by a king and queen and takes up residence in sea sponges, where soldier shrimp stand guard, weapons in hand.

I've hitherto presented a view of nature that's less than pleasant, that's all just murder and tongue-eating and zombifying and whatnot. And that's because it *is* less than pleasant out there. While Mother Nature solves problems, in so doing she necessarily creates problems, and thus does the push and pull go on and on, with critters evolving weapons and their prey evolving defenses. Really, amid all that, in the natural world you'd be lucky to die of old age. That's a crappy situation and it sucks and I'm sorry, but as Alfred, Lord Tennyson, rightly put it, nature is red in tooth and claw. So it's no wonder that some creatures have found fit to band together. After all, as the old cliché goes, there's safety in numbers (this is a neat little writer's trick, by the way—if you admit you're using a cliché, you don't get dinged for using clichés).

And there are few gatherings more peculiar than those of the highly social pistol shrimp, which, as their name suggests, have evolved some of the most powerful weapons on Earth. Individually, they're intimidating, but as a group they make up armies and establish forts other creatures would be damned fools to try to conquer: These creatures form the sea's only monarchies, on par with the gatherings of ants on land. And their collective gunfire is deafening.

The weapon responsible for all the racket is the pistol shrimp's enormous, grotesque-looking claw (only one—its other is a smaller pincer), which in some species can grow to half the length of the animal's body. By contracting certain muscles, the shrimp brings back one half of the snapper, which has a protruding bit known as a plunger, into a locked position while the second half, which has a socket, remains immobile. When the shrimp contracts another muscle, the two halves slam together, with the plunger striking the socket with so much force that water flies out of it at 105 feet per second. The impact forms what are known as cavitation bubbles, and when they collapse, they

"THE CACKLING AND CLUCKING OF A BARNYARD FULL OF CHICKENS"

Pistol shrimp make such a racket that they've been known to disrupt sonar equipment. In World War II, a constant crackling noise nonplussed one American submarine skipper in Indonesia, a din he chalked up to the Japanese having "some newfangled gadget that they drop." As the *Milwaukee Journal* recounted in 1956 with enviable and perhaps unnecessary thoroughness: "The sounds were described in terms that were both poetic and harshly blunt: Such as coal rolling down a chute, fat frying in a pan, the dragging of heavy chains, croaking, moaning, whining, grunting, drumming, a subdued steamboat whistle, the rasping of a saw on a strip of steel, the cackling and clucking of a barnyard full of chickens, the 'put put' of a poorly running outboard motor, a badly hurt and groaning man, and so on."

heat the surrounding water to 8,000 degrees Fahrenheit and send out a shock wave so powerful that it can instantly kill prey. Strangely, it's not the impact of the claw itself that makes the noise, but the violent collapse of these bubbles.

In subtropical shallow waters, the shrimps' collective snapping produces much of the ocean's ambient noise, a sort of burning-twig crackle. And the sound—my God, the sound. A single snap can hit 210 decibels. For perspective, the American Speech-Language-Hearing Association considers a lawn mower's 106 decibels to be "extremely loud," and a firecracker's 150 decibels to be "painful." That's as high as they bother going. The pistol shrimp's snap is literally off the charts. (Another writer's trick: If it's literal, it's not a cliché.)

Now, we humans use guns to kill things and maybe sometimes fire in the air when we're really excited about something, as does the pistol shrimp with its own weapon (the killing-things bit, not the other part). But for the shrimp, a gun is so much more: They'll use snaps to communicate, and in fact they have sensory hairs all along their pistols that help them detect the shock waves. The shrimp also use them to fight, not necessarily maiming each other but instead just kind of blasting off warnings. Some species can even excavate rock to form a home with successive blasts of cavitation bubbles.

But back to the monarchies. A handful of these pistol shrimp species form what are known as eusocial societies inside sea sponges. And that's really, really odd. Such organization is expected for things like ants and termites (but not bees, as it happens—most of their species are solitary), but this is unheard of in the oceans. The society is all structured around a queen, who's far larger than any of her subjects. After all, the bigger a female is, the more young she can bear. She's the only one who mates—with her king, of course—ruling several generations of offspring, like a bossy grandmother of sorts. And while she towers over her underlings, she lacks a snapping claw, suggesting that she relies on the colony's soldiers for protection.

HERE, HAVE THIS TOKEN OF EVERLASTING ENTRAPMENT

Another intriguing species of shrimp takes up residence in the Venus' flower basket sponge, which looks like a long tube of glassy mesh. These so-called wedding shrimp enter as a young pair, eventually growing too large to exit through the sponge's lattice. So they'll mate and spend the rest of their lives trapped inside. In Japan, a traditional wedding gift is the flower basket sponge, complete with its expired inhabitants. If there's a better metaphor for marriage than that, I don't know it.

While these warriors may be small, they have the advantage of numbers. And for pistol shrimp, the claw isn't only for intimidating each other. It's an alarm. Should one of the soldiers come across an intruder trying to weasel into the sponge, be it another species of pistol shrimp or something like a fish with a mind to wipe out the colony, it'll snap a rhythm to call in reinforcements. And when they arrive, they'll all join in on the rhythmic snaps to drive the enemy away. It appears to be quite effective, in particular when driving off other species of pistol shrimp. They'll also do this to fend off the sponge's own enemies, things like sea slugs, thus paying their rent.

Where this gets a bit complicated is that in the animal kingdom, selfishness typically rules. Individuals are in it to pass down their own genes at whatever cost, not risk their lives for anyone else. So if you're a eusocial pistol shrimp, and you can't mate, why on earth would you stick around to protect your queen and siblings? It would seem to defy Darwin's idea of evolution by natural selection.

The answer is something called kin selection. You don't have to personally pass down your genes to pass down your genes. If you can help guarantee the survival of your kin, in a roundabout way your own blood will make it into the next generation. Yeah, you don't get to have sex, that's a bummer. But hell, if you're a social pistol shrimp you've got a sweet pad and a lovely family. What more could a crustacean ask for?

Sociable Weaver

PROBLEM: The desert is somewhat . . . toasty, not to mention full of snakes.

SOLUTION: A nest might do nicely. But even better? The biggest bird nests on the planet, climate-controlled structures so big they pull down trees.

Squatters, mutters a bird known as a sociable weaver. *Bums. Miscreants. All of them. Invading anuses and mouths and sponges. Try building something for a change.* And off the busy little bird flies to grab yet another twig to help construct the most incredible nest on Earth—hands down. I'm talking twenty feet long. Thirteen feet wide. Seven feet thick. All told, it's two thousand pounds of nest that's home to up to five hundred birds living in one hundred chambers. Really, it's not a nest. It's a mansion so big that it'll sometimes topple the tree it's built in. But should the nest avoid that fate, it can stand for a century in the plains of southern Africa, serving as a home for generation after generation of sociable weavers.

There are no queens here to boss everyone around. This is an anarcho-syndicalist commune, baby, through and through. The birds work together, scouring the landscape for twigs to weave into the massive structure, as well as grass to line their chambers, which actually aren't interconnected. Instead, each has a single entrance at the bottom of the nest. Situating the chambers like so helps keep the rain out and makes it that much harder for snakes and birds of prey to infiltrate the structure. Interestingly, though, while the sociable weavers will team up to chase away invading pygmy falcons, at times they'll let the predators move in. Sure, the falcons might pick off a weaver from time to time, but keeping them around has its benefits, too: They prey on snakes. Think of it as a protection racket, however fragile the alliance may be.

As with pistol shrimp society, what seems to hold the weaver group together is kin selection. By working cooperatively to construct such a huge nest, the weavers have advantages over birds that individually build small nests. With such a mansion, weavers enjoy excellent insulation in addition to protection from predators. Nighttime temperatures in their habitat can plummet to below freezing, but by huddling together in their chambers, they can crank up the thermostat to a toasty 70 degrees Fahrenheit. And when things heat up during the day, the nest retains the coolness of the previous night.

WAITER, THERE'S SPIT IN MY SOUP

Asia has a bird whose nest may be dwarfed by the sociable weaver's, but it's far more valuable. The edible-nest swiftlet constructs a treasured little home that sells for a thousand dollars a pound in China, where it's made into soup that's said to have medicinal properties. Only a chucklehead would pay a grand for a bundle of sticks, though, right? These turn out to be very special structures: Instead of building with sticks or grass like other birds, the swiftlet uses its sticky spit to spin a cup-shaped nest on cave walls. The nests are so sought after that in Indonesia, folks are spending as much as sixteen thousand dollars to build three-story concrete nurseries for the swiftlets. The creatures come for the swiftlet calls booming out of CD players, and stay for the food and climate control via a network of misters.

Here's the problem, though. Unlike the pistol shrimp, all of these sociable weavers are plenty capable of breeding. And while it makes sense to work cooperatively with your family for the chance to pass down your shared genes, first and foremost you need to think about yourself. While a weaver's relatives are all flying around, busting their butts to build the structure as a whole, it's mighty tempting for an individual to instead invest time in maintaining its own chamber, boosting the chances of its own survival and that of its young. Or a bird could slack off entirely, focusing on foraging so it can grow big and strong, or mating so it can pass down even more of its genes. Thus we'd expect more of these selfish genes to show up in the population, since there must necessarily be more offspring of self-centered individuals. And indeed, among the sociable weaver colony there are the cheaters. Some sneaky chumps think they can get away with not pitching in, enjoying the benefits of the nest without contributing their fair share.

But maybe they can't for long. In captivity, it seems enforcers have formed gangs to pursue a slacker, catching it and giving it a good pecking. We should, of course, be somewhat wary of these accounts, though. Animal behavior changes dramatically in captivity, plus the sociable weavers in this particular case were in a small enclosure with little room to flee. But out in the wild, scientists have observed plenty of chases, which could serve to keep the slackers in line. The birds aren't getting evicted, mind you, just getting a reminder of their place in the society. And after spending some time away, they return and resume work on the nest. The aggression is bad enough that the chance of injury becomes too great, outweighing the costs required for the birds to contribute

AND NOW, FOR MY NEXT PARTY TRICK . . .

Lacking the sociable weaver's safety of a community, a bird known as the fulmar has evolved a rather more inventive way to defend itself: Its chicks will vomit a fetid oil on predatory birds up to ten feet away. But this is more than a wretched inconvenience. That oil won't come off, and this is a big problem for a bird that needs its feathers to remain clean in order to stay waterproof. The fulmar's victims can actually perish after a good coating of puke.

But leave it to humans to make the best of a bad situation. In the 1800s, the people of the Scottish islands of St. Kilda sold fulmar feathers as stuffing for mattresses and pillows, after fumigating them, of course. The feathers were said to keep lice and bedbugs away, yet, sure enough, after three years the smell would return, to the horror of the sleeper. At that point you pretty much had to burn your house to the ground.

their fair share. Sure, they may still at times selfishly focus on their own chambers to help guarantee their own survival, and therefore the continuation of their genetic lines, but they'll also contribute to the nest at large to ensure the success of the society as a whole, and therefore the continuation of their genetic lines by way of their kin.

It sounds like pure anarchy, I know, and I promised you an anarcho-syndicalist commune. But there actually seems to be a cryptic signal of order right below the sociable weaver's beak: a bib. Extending down their necks are bands of black that contrast with their otherwise cream-colored feathers. It's what's known as a badge of status. The longer the bib is, the higher the bird is ranked and the more dominant the bird is toward its peers, though individuals at the top don't find themselves challenged much at all. It's so effective that overall size, a popular marker of dominance in the animal kingdom, isn't as important to the bird's status as is the size of its bib.

Thus it seems that order in social weaver society comes by both the bib and the chase. And by sticking together and overriding that pesky evolutionary proclivity for selfishness, sociable weavers have a distinct advantage out there on the plains of southern Africa. Offering a clue to how this sociability evolved in the first place are the nests themselves. Because the chambers are not interconnected, this suggests that sociable weavers, like American settlers circling the wagons, once built smaller nests near each other, perhaps to enjoy the safety that comes with numbers (sending out alarms, forming gangs to chase away predators, etc.). Over time they merged the individual nests into some of the most magnificent homes on Earth and, with selfishness largely overruled, the sociable weaver came to rule the savanna. All it took was a good old public beating now and then.

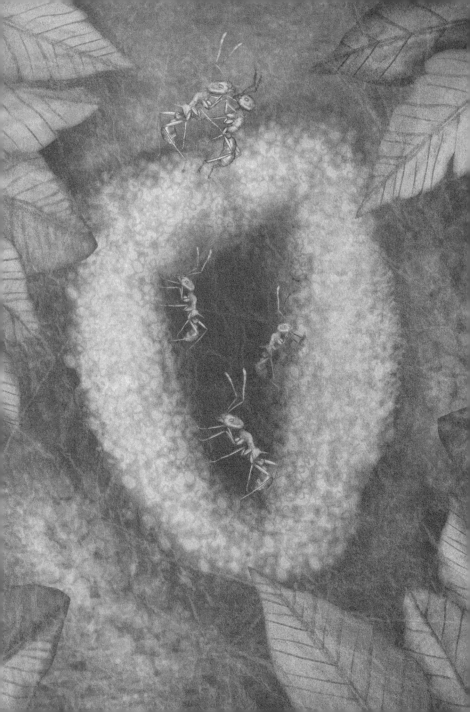

Hero Ant

PROBLEM: Rain forests are wet and packed with disagreeable organisms.

SOLUTION: The so-called hero ant creates what is surely the simplest, yet most perplexing ant nest on Earth. And defends it—kamikaze-style.

Oh, how the pursuits of natural history have changed over the centuries. Back in the day, Maria Sibylla Merian and Charles Darwin and Alfred Russel Wallace were prone to wandering, sometimes for years. After all, chances were the maps that could help them didn't yet exist. And perhaps that was all part of the romanticism of early natural history: There was as much an aim to unveil new species as there was a wanderlust that sometimes involved falling victim to a caterpillar. But as science matured and grew increasingly complex, natural history's adherents had to fragment into specialties: Entomologists love their bugs, ornithologists their birds, ichthyologists their fish. The wandering generalist grew endangered, and not only because a good amount of them died in the process. These days, cash-strapped scientists are on expensive, and therefore necessarily precise, missions.

And so in 2013, Brian Fisher, one of the world's foremost experts in ants, found himself in the remote rain forests of Madagascar, carrying the ultimate luxury of the modern field biologist: the iPad. He didn't have the resources that Darwin had (thanks to dear old rich Dr. Papa Darwin, who bankrolled his son's adventures—perhaps reluctantly, but that's a whole different story) or the luxury of time that Merian and Wallace had. What he did have, however, were thirty porters to deal with, porters who hadn't the slightest idea where they were headed. Hence the iPad, loaded with the latest imagery from an imaging company that had agreed to train its satellite on a mountaintop near where Fisher had, on an earlier expedition, collected the mysterious ear ant—soon to be given the honorific of hero ant— whose nests jutted out of the sides of cliffs as earlike funnels. He and his colleagues and the porters were there to figure out why exactly an ant nest would appear to have a megaphone facing out to the world.

NEED TO PASS THE TIME SO YOU DON'T DIE WHILE SCUBA DIVING? THERE'S AN APP FOR THAT (ALSO A CLICHÉ, BUT REMEMBER OUR RULE)

I once interviewed a pair of marine biologists who regularly scuba dive four hundred feet deep, a ridiculous figure if survival is among your interests. That's so deep that of the seven total hours of diving time, they can spend only twenty minutes down at their destination because they have to factor in the time spent ascending and decompressing to avoid the bends. They make a whole bunch of calculated stops coming up, which begin at a minute long, culminating in ones near the top that last two hours. So how do they stave off insanity? By playing Angry Birds on a waterproofed iPad, which I imagine they could even write off on their taxes. Well played, scientists.

The team was working off three theories that could explain the perplexing domicile. For one, the funnel might serve as an obstacle for any number of small marauding predators in the rain forest—the hero ants may post up at the edge of the structure and fight the invaders from an elevated position. Two, it could be an antiflooding measure. This is, after all, called a rain forest on account of being a forest where it rains a lot. Third, the funnel could help facilitate air flow, for a shape like that would presumably work much better than a simple hole for ferrying oxygen in.

Now, I take it you're imagining quite the network of tunnels connected to that megaphone entrance. But this isn't that kind of ant. The funnel itself is less than two inches wide, and it opens into a single chamber that's only three inches deep. That's all. Here you'll find the queen and her young and all the workers shuffling in and out. It's the only known ant on Earth that does this, but such puzzling architecture is likely a product of the material in which the ant makes its home.

It turns out that the wet clay of the walls isn't conducive to gas exchange between the atmosphere and the nest's interior, so there's a real danger of carbon dioxide accumulation inside. The nest is probably so shallow because it has to be: It'd be hard to get circulation going in meandering tunnels. And what Fisher and his team calculated was that by building out that funnel, the ants boost the air flow into the nest by six times, no small matter when their home only has one air duct. The team also found that the funnel helps keep out water. A simple tube jutting out from the nest, though, would have had the same benefit. Perhaps it was that the funnel-building behavior evolved to facilitate air flow, and had the bonus of keeping the nest good and dry.

So what about the tunnel aiding in defense? Well, that's where a field biologist's job gets fun. The team placed eight other ant species next to hero ant nests, including individuals from other hero colonies and two species with an appetite for the larvae of other ants, and observed how the invaders navigated the funnel and how the unwitting hosts reacted. It turned out the funnel mattered . . . not one iota. The hero ants didn't sit there at the entrance and defend, and every single one of the other ant species had no problem scrambling into the nest. And even then, they didn't immediately set off any alarms. The heroes got worked up only when they directly ran into the invaders.

And, boy, did the heroes get worked up. They didn't swarm, like army ants so famously do. Instead, hero ants turned out to be one-on-one scrappers. Fisher and his colleagues watched plucky individual hero ants not just snag an intruder, but drag it out to the edge of the funnel and then—leap. Locked in combat, the pair would tumble off the cliff, and thus did the lowly "ear" ant become the "hero" ant. Both parties could survive the impact, since they're so light and have such strong exoskeletons. Inevitably, the hero

FURTHER ADVENTURES IN WATCHING ANTS FALL OFF THINGS

If we might call the hero ant's defense a kind of intentional chaos, the ant species of the *Cephalotes* genus are surely the elegant skydivers of the ant world. These are tree dwellers, and when threatened they will leap off a branch, legs splayed. With a few deft movements of their limbs, they can steer themselves right back to the trunk of the tree and make their way up to the colony again. We know this because a scientist once took the time to sit in a tree and watch them— but not before picking them up and painting them white and tossing them off himself. I mean, it's okay, really, because 95 percent of the ants ended up landing back on the trunk. As for that other 5 percent? Well, no one ever said skydiving doesn't come with risks.

would make its way back up to the nest, on account of these being its stomping grounds, but the odds of the intruder doing the same to try its luck a second time were slim. And the ant's return is a welcome one: It's truly a precious resource for the colony's queen, who's ruling a society of as few as a dozen individuals.

Now, the classical high-school-biology notion of ant sex is that winged males, whose only purpose is to reproduce, and queens from a given neighborhood's colonies take flight and mate. Afterward the males die and the females land somewhere, break off their wings, and start a new colony from scratch.

But not with the hero ant. The hero ant queen, known rather unflatteringly as an ergatoid, has no wings and appears to wait for the winged male to come to her. It isn't yet entirely clear whether it happens when she's in her original colony or when she's moved to a new one, but it doesn't much matter for her. The important bit is that the queen always takes a squad of workers, which are all female, with her when she packs up and leaves, meaning she's got a workforce to build a new home wherever she sees fit. In winged species of ant, the queen has to wait to have her young before the workforce really spins up. So, however precarious the situation of the hero ant colony may seem, the society always has a pretty good start at things. It's just that a tumble or two comes as part of the territory.

CHAPTER 4

You Live in a Crummy Neighborhood

In Which Tiny Gummy Bears Go to Space and Zombie Ants Stagger Through the Rain Forest

Sometimes even the finest home in the finest mouth or anus simply won't do: Some neighborhoods are just terrible. So animals have a couple options. They can move away, or they can adapt. A spider sick of getting eaten, for instance, could move into the water. Or if you're the tunnel-dwelling naked mole rat, you could evolve stretchy skin that helps you squeeze through the passages, skin that may hold the key to curing cancer.

Water Bear

PROBLEM: Mud is a lovely place to live—as long as it stays mud.

SOLUTION: Should its muddy home start to dry up, a tiny critter known as the water bear can dry out almost completely for up to thirty years, only to reanimate once it hits water again.

Life doesn't like being told no. Even in the deepest depths of our oceans, where frigid temperatures and crushing pressure and scarce food combine to make for an unpleasant habitat, life endures. Everywhere you'd figure to be devoid of life has something holding on: Wildlife proliferates in the contaminated forests of Chernobyl while bacteria kick back in the scalding hot springs of Yellowstone. Only in the vacuum of space should we expect sterility, for surely nothing could survive there.

Well, as it turns out, there's one tiny critter that especially doesn't like being told no.

In 2007, the European Space Agency launched a satellite with very special passengers indeed: two species of tardigrade. Also known as water bears—on account of resembling bears, or, even more so, gummy bears, only with twice as many legs and not nearly the same taste—these microscopic invertebrates had already grown legendary for their resistance to heat and cold and radiation. So some scientists got an idea: Why not fire them into space? And so the scientists did. Once these water bears reached low Earth orbit, they were exposed to the vacuum for ten days, then brought back home. Incredibly, they survived. Like, really well. Not one or two that came back coughing and moaning, but a whole slew of them. Their eggs even survived and hatched into healthy young.

Here's the problem, though. The water bears cheated. Don't get me wrong, surviving space is a feat that boggles the mind. But they have an evolutionary ploy. You see, while these creatures live in dirt and moss and such, they need those materials to be at least a little bit moist. But should their habitat start drying out, water bears can dehydrate down to 3 percent of their typical water content, folding in on themselves and entering what's known as cryp-

THE ADVENTURES OF MUTTNIK: THE MOST RUSSIAN TALE IN THE HISTORY OF RUSSIA

There's a long, exceedingly weird history of sending animals into space, and leave it to the Russians to provide most of the weirdness. The first animal to enter orbit was a stray dog recruited from the streets of Moscow (the thinking went that strays were scrappier and better cut out for surviving space) only nine days before her rocket launched in 1957. But there was a glaring hitch: No one had had time to think up a plan for reentry. It would therefore be a one-way trip for the pooch. So before launch, a Soviet scientist took her home to play with his kids. "I wanted to do something nice for her," he recounted. "She had so little time left to live."

The pooch never made it to reentry—she overheated and perished a few hours after launch. But she's now immortalized in a Moscow monument, standing confidently on a rocket. The Russians called her Laika, but in America, she'll forever be known as "Muttnik."

tobiosis, essentially suspended animation. (Cryptobiosis was first discovered in a different creature way back in 1702 by a scientist named Antony van Leeuwenhoek, who collected dried sediment from house gutters and added water. The material erupted with life—"animalcules," as they were known then, a blanket term for itty-bitty creatures.)

It's in cryptobiosis that water bears are virtually indestructible, as their bodies mobilize a sugar that replaces the lost water and protects their cells. They can stay like this for at least thirty years and still reanimate fine. So, that said, when the scientists sent their water bears into space, they were dried out, only to be reanimated back in the lab, a process that takes about an hour.

I say *virtually* indestructible, though you may as well say indestructible. Short of throwing them in a bonfire, there isn't a good way to kill water bears. Water boils at 212 degrees Fahrenheit, but that's like a nice day at the spa for these animals: They can turn the thermostat up to 300. Try boiling them in alcohol, you say? Won't work either. Water bears can withstand pressures six times higher than in the deepest oceans, and you can bombard them with hundreds of times the radiation that would kill a human, and they shrug it off. In fact, when they were hanging out in space, some even got a dose of raw UV radiation, an extremely destructive energy when you don't have the luxury of an atmosphere protecting you. Almost all perished . . . but a handful still survived.

Perhaps most incredible, though, is the water bear's resistance to cold. Absolute zero, −459.67 degrees Fahrenheit, is the lowest temperature possible. Well, technically nothing can ever get that cold, but in the lab, scientists can get close, the world record being one ten-billionth of a degree above absolute zero. When in cryptobiosis, the water bear can survive temperatures down to 458 degrees F, a degree and a half away from absolute zero. That's even lower than the lowest temperature possible outside of a lab, about −455 F, which you'd only find out in the darkest, loneliest depths of the universe.

Let that sink in. The water bear is capable of surviving temperatures no organism would ever experience in the known universe, save for in a mad scientist's laboratory. And that has startling implications for our conception of life. As a denizen of Earth, the water bear has it relatively easy. It's got a nice atmosphere and lots of water and sunshine. But of course not all planetary bodies are so hospitable. Our own moon, for instance, has temperatures that fluctuate between −280 and 260 degrees F, easily tolerable for the water bear. There's also all kinds of nasty radiation out there, which can scramble DNA. That's a bit of an issue if you were hoping to pass down your genes—well, it's a problem for most life as we know it, but not the water bear. It can repair DNA damaged by radiation. And if the water bear is so hardy, there must be other creatures out in the universe that are even hardier, inhabiting all kinds of weird, brutal environments we'd assume to be sterile. So looking for life elsewhere in space isn't necessarily about finding another Earth. The water bear shows that life can take hold where you'd least expect it.

It's because all a water bear ever wanted was to survive a dry spell. As an added bonus, when they dry out, their husks can get caught up in the wind, which spreads them far and wide, so there aren't many places you *won't* find a water bear or two, from the deep sea on up to the tops of mountains. So I don't know about you, but I sure am proud to share a planet with the water bear. Hell, I say we put them in rockets and fire them out into space to colonize other worlds so aliens can enjoy them as well. Or is that wildly irresponsible? That's wildly irresponsible, isn't it.

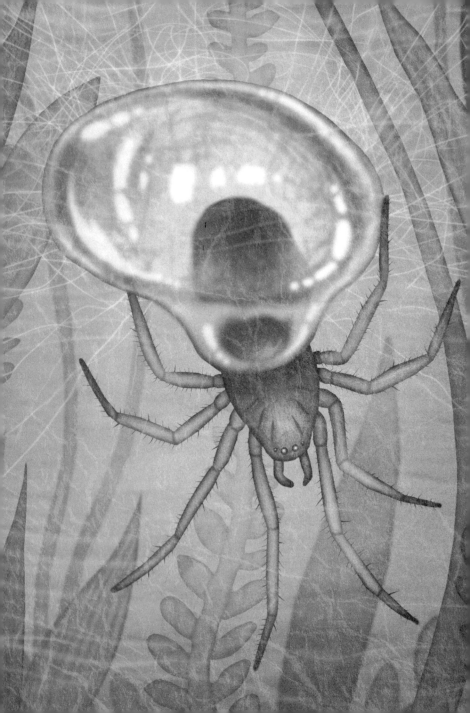

Diving Bell Spider

PROBLEM: Life on land is full of all kinds of nasty predators, as well as competition for food.

SOLUTION: A certain spider leaves it all behind and goes aquatic, using its butt to collect a bubble of oxygen and start living in the water . . . permanently. This is the only spider that lives its entire life underwater.

Life on land is no picnic. Well, it is a picnic if you happen to be near the top of the food chain. But if you're toward the bottom, you *are* the picnic. Take the spiders, for instance. They're master hunters in their own right, but all manner of other critters have them on the menu. So a spider looking to survive can hide and ambush its prey, like the trap-door spiders do, or mimic an unpalatable species, like a ladybug spider does, with its red-with-black-spots outfit (yes, ladybugs are toxic). Or you can take a page from the diving bell spider's book and pack up and leave land altogether.

I don't want to tell spiders how to live their lives, but they really don't belong in water. There's the whole problem of, you know, breathing and stuff. But for that the diving bell spider has an incredible solution. Its body is coated in hydrophobic hairs, so as the spider is swimming around it'll periodically poke its bum above the surface, trapping a silvery bubble of air that it carries around with it underwater.

I know what you're thinking. *Another creature that breathes through its butt? What's with this guy and animals that breathe through their butts?* Well, I have good news, because spiders aren't in this club. They do, however, breathe through their abdomen. The diving bell spider does it two ways. First, slits in the exoskeleton open into what are known as book lungs, so called because they have a series of plates, filled with hemolymph (the arachnid version of blood), that draw oxygen from the air and look like pages of a book. And second, holes in the exoskeleton allow oxygen to flow directly to organs and other tissues. With these two systems, the diving bell spider just has to keep air around its abdomen, leaving the mouth free to feed.

This system comes in handy when the spider builds its namesake home. Swimming around all the time is dangerous and energy intensive, so the spider spins shelter between vegetation. Yet it doesn't build out an expansive web like its terrestrial cousins, but instead a more spherical, hollow one—a bell. The spider need only make it big enough to insert its abdomen—though sometimes

BUTTS: AN UNDERAPPRECIATED RESOURCE

I once had an editor somewhat seriously suggest that I seek therapy for my proclivity for writing about animals that do weird things with their anuses. But it's not my fault, really. It's that a lot of critters do weird things with their anuses or, in the case of the pearlfish, with *other* critters' anuses. I think this speaks to the larger issue of our anthropomorphization of the animal kingdom. Is it weird to us that pearlfish swim up sea cucumber butts and that there's a species of leech that feeds only on the rectums of hippos? Sure, but the natural world has been humming along with such "eccentricities" for millennia. Or maybe I'm trying to justify not spending a bunch of money on therapy. Ironically enough, though, only a therapist could tell me that.

it'll expand the nest to become big enough to move freely in—and it takes care to run lines to the surface that it'll clamber up to grab more air and deposit in the bell. Thus can the diving bell spider live underwater indefinitely. And thanks to the laws of physics, it doesn't even need to surface that often, maybe as little as once a day if it isn't active much. This is because, just like at the surface of the water, oxygen is exchanged between the air and water through the web, so the oxygen the spider consumes is readily replaced.

Now, male and female diving bell spiders go about life a little differently. Males are far more active than females, which tend to hang out in their bells all day (they'll also raise their young here, enlarging the bell as the youngsters grow). And believe it or not, these spiders are decent swimmers, especially the males, rapidly paddling their legs to propel themselves in pursuit of small fish and crustaceans like water fleas. The females, though, prefer to do their hunting at home, waiting for something to bump up against their webs, then scurrying out and attacking. But before they bring their prey in, they'll enlarge their bells and grab more air. With a bit more room, they can then settle in to feed.

These lifestyle differences could explain why males are bigger than females. That's a bit weird, because as I mentioned earlier, such a size discrepancy simply isn't the norm in the animal kingdom. But for spiders, this is downright goofy. In very few species do the boys grow bigger than the girls, because while females need to be big enough to bear a lot of young, male spiders focus on mobility. For spiders on land, being small helps them to get around: Some species, in fact, are so small that they can fly around by sending out a line of silk that the wind picks up, a trick known as

SOME SERIOUS MOMMY ISSUES

Apart from the sexual organs, physical difference between the sexes of any given species is known as sexual dimorphism. This can be something like size, as in the diving bell spider, or ornamentation, as in the peacock and peahen. Perhaps the most astounding sexual dimorphism around is that of the *Strepsiptera* insects, nasty little things that invade the bodies of other bugs. A male looks fairly normal, almost like a fly, with wings and legs and all that. But a female is a glorified bag of eggs, with no eyes or legs or wings or even mouthparts. She invades a host and pokes her oviduct out of its abdomen, and a male comes along and fertilizes her. The young—up to a million (not a typo)—consume their mother from the inside out before erupting out of her and the host and into the world.

ballooning. It's so effective that sailors hundreds of miles out to sea have reported seeing flying spiders landing on their ships . . . so maybe it can be *too* effective.

But when it comes to mobility in an aquatic ecosystem, it pays to be *bigger* so you can more easily cut through the resistance of the water. Thus larger male diving bell spiders—which also have elongated front legs to aid in paddling—have a few advantages. For one, the better they can swim, the more likely they are to snag prey and avoid becoming lunch themselves. And keep in mind that dragging a bubble down from the surface is quite difficult, since the spider becomes much more buoyant when it's carrying one. Size, then, equals power. Plus, the biggest, strongest swimmers will get access to more females. Hence natural selection favors larger male diving bell spiders, the opposite of their counterparts on land.

Sure, moving into the water comes with its challenges. There are predators just as there are on land, and you have to somehow develop the ability to breathe underwater. But setting up shop in the new neighborhood also comes with a very important perk: The spiders get themselves a nice little niche, sitting there picking off those crustaceans. Diving bell spiders have left their fellow arachnids—their stiff competition—behind, and are now basking in an underwater monopoly. Think of it like moving from the mean streets into a secluded cabin. Only quite a bit wetter, I suppose.

Zombie Ant

PROBLEM: A fungus in a windless rain forest is going to have trouble getting its spores around.

SOLUTION: *Ophiocordyceps* invades ants' brains and mind-controls them up into trees to very specific spots, ordering the zombies to bite down on leaves before killing them. The fungus then erupts out of their heads and rains spores on their comrades below.

You're not going to buy a word of this. But here goes.

The fungus spore begins by sticking to a carpenter ant's exoskeleton, using enzymes to rot away the cuticle, all the while building up pressure to explode itself into the ant's body. Here it reproduces over the next three weeks so prolifically that half the ant's weight ends up being fungus. Meanwhile, the zombie ant is acting normally, shuffling around doing whatever it is ants choose to do with their day, while its comrades remain none the wiser to the fiend in their midst.

Then the ant disappears. The fungus has ordered it to flee the colony around noon—*always around noon*. And it drives the stumbling and clearly unwell ant to a leaf ten inches off the ground—*always ten inches off the ground*. The fungus has found the perfect humidity and temperature with which to grow and, critically, it's positioned itself right near one of the colony's well-worn trails. The fungus orders the ant onto the underside of the leaf, where the zombie drives its mandibles into the vein, holds tight, and perishes. And with that, the fungus erupts out of the back of the ant's head as a stalk, raining spores onto the trail below. Even if the fungus has steered the ant close to the path, but not directly over it, the stalk will grow at an angle that allows it to arch the stream of spores onto other ants on the trail. And multiple zombified ants can end up attacking a given trail, creating a snipers' alley of sorts, picking off their erstwhile comrades and beginning the whole impossible freak show all over again.

I've got this funny feeling you still don't believe me, that the fungus, *Ophiocordyceps*, can orchestrate the world's most startlingly complex parasite-host relationship, even though it doesn't have a brain of its own. (The fungus is alone in this book in that it isn't quite an animal and isn't quite a plant, yet it behaves with the shrewdest calculations of the cleverest animals to create a unique creature: the zombie ant.) Or that the various species of the fungus specialize in attacking their own particular species of ant. But it's

---

GROWTH INDUSTRY

While carpenter ant colonies clash with mind-controlling fungi, the various species of South American leaf-cutter ants depend on another group of fungi to survive. Leaves in the tropics can be quite toxic, so these foraging ants don't eat them. Instead, they saw off bits and carry them back to base. Here, other members of the colony chew the leaves up and spit them out for a fungus to dine on, and the ants in turn dine on the proliferating fungus. The problem, though, is that the promotion of fungal growth can let in a hyperaggressive mold. But not to worry: These ants are coated in special bacteria that check the invader's growth, allowing the beneficial fungus to take hold and continue providing the colony with food, free of competition.

■ THE WASP THAT BRAINWASHED THE CATERPILLAR

happening. Every single day. And it's been happening for a long, long time—scientists have found leaves 48 million years old that have the telltale bite marks of zombified ants.

Somehow the mindless fungus has figured out how to use ants as vehicles, likely because of a lack of wind in the forest to transport the spores. There's too much dense vegetation. So in order to spread around, *Ophio* has evolved with the ants for millions upon millions of years—exploiting them, mind-controlling them—to distribute itself around the jungle. Really, that's all quite embarrassing for the ants, which have built complex societies, only to fall victim to a mindless fungus. (Consider as an analogy your houseplants brainwashing you.)

Scientists are only beginning to understand what's happening to the ant's body and mind during all of this, but we can say with reasonable confidence that it isn't pleasant. What's clear is that the fungus is producing neuromodulators, compounds that monkey with the ant's neurons, which are of course responsible for coordinating movement. The fungus's control, though, isn't seamless. As the ant makes its way out of the colony, it convulses and collapses from time to time as the neuromodulators degrade its muscles. A lot like the way booze affects the human brain and coordination, so too does the fungus manipulate the ant, only much, much more precisely. The bug sways around as the fungus pulls the strings, guiding it to that exact spot where the stalk has the best chance of growing.

Intriguingly, *Ophio* is related to ergot, the fungus that gave humanity LSD. Both fungi clearly have psychoactive properties, and both have found their way into the human psyche. On the Tibetan Plateau, for instance, there are related species of *Ophio* that attack different insects, though they don't go through the sophisticated life cycle of the ant-zombifying varieties. And around 1,500 years ago some human caught on, noticing that yaks were gobbling up grass and some side dishes: ghost moth caterpillars with fungus stalks erupting out of their heads. The yaks would trip out, running around like a bunch of idiots. And just like that, a trade was born. These days the fungal caterpillars can sell for two thousand dollars an ounce and are said to make you good at sex.

The Tibetan variety of *Ophio* is after caterpillars, but the ant-hunting variety faces a whole new slew of challenges with its host. You've probably been wondering why the hell the fungus would go through all the trouble to steer the ant around the rain

THE *OPHIOCORDYCEPS* ZOMBIE ANT FARM ACTION SET FOR SADISTIC CHILDREN OF ALL AGES

For a long while scientists thought that mind-controlling fungi were more of a rain forest thing, but in 2009 the world expert on *Ophio*, David Hughes of Penn State, came across photos from a South Carolina woman who'd found them in her backyard. This variety is just as clever as its cousins, perhaps even more so. In the rain forest, leaves are there to bite onto year-round, but in South Carolina, there are of course seasons that strip trees of their foliage. As such, this North American fungus has evolved to command the ant to bite onto stems instead of leaves, since stems are available regardless of season. So take note: If you're an ant you're not safe in many places. Except maybe in an ant farm. That is, until someone comes up with the *Ophiocordyceps* Zombie Ant Farm Action Set for Sadistic Children of All Ages.

Now if you'll excuse me, I have a fortune to make.

forest when it has already infiltrated the colony, where there are plenty of other fresh bodies to corrupt. Well, among the many brilliant strategies of ants is their keen eye for misbehavior. If one of their own starts acting strange, they'll drag it out of the colony and dump it in a nearby graveyard, because more than likely the thing has fallen victim to some kind of disease or, in our case, a homicidal fungus. This is known as social immunity, and it's how ants avoid outbreaks that can wipe out whole colonies. But ant-exploiting parasites like *Ophio* have evolved alongside their hosts for millions of years and have come up with clever solutions. The same goes for the ant-decapitating fly. It knows that there's no sense in sticking around the colony to get spotted. It steers its victim far away, where it can destroy the ant in peace.

Are you still with me? Is this in any way believable for you? All I can do is insist that it's happening, and that there are still many questions here. Why, for instance, doesn't the fungus eventually wipe out the colony, leaving itself without vehicles to get to other ants? Part of the answer might be that these parasitic fungi have other parasitic fungi that exploit them, castrating the stalks after they erupt from the ants' heads. But what about when something catastrophic happens, like outside forces decimating the colony? Can the fungus make the leap to another colony? Scientists still have much to learn, but we can say with confidence that ant zombification has been around for millions of years, with random mutation after random mutation allowing the fungus to assume a morbid mind control over its host.

Pink Fairy
Armadillo

PROBLEM: Living in the desert means enduring brutally high and brutally low temperatures.

SOLUTION: One burrowing armadillo has turned its protective shell into a blood-vessel-packed radiator that gives off extra heat or absorbs heat when needed.

One of my biggest pet peeves is the characterization of the desert as "lifeless." Nothing could be further from the truth. The desert is crawling with creatures—it's just that so many of them, from insects to rodents to bats, only emerge at night, when the climate is bearable. But around the clock there's a secret world that's bustling beneath the apparent calm: that of the burrowers. And there's no burrower more endearing, more charismatic, or more remarkable than the pink fairy armadillo of Argentina, one of the rarest mammals on Earth.

Describing the pink fairy armadillo is an exercise in absurdity. The creature is cylindrical and can fit in the palm of your hand, sporting enormous claws that it uses to dig through the soil. Its body is covered with an almost perfectly white fur, topped with a pink band of a shell that stretches along its back, from its nose to its rump, which is flattened so as to tamp back soil as it burrows. All in all, it looks more like a torpedo than an armadillo.

OUT ON A LIMB

A couple thousand miles away from the pink fairy armadillo, in the deserts of Baja California, is another burrower that has also evolved a unique body shape to better move through the ground. Known as the Mexican mole lizard, it's not actually a lizard but instead part of a unique group, amphisbaenia. The critter is long and thin like a snake, but has two front limbs it uses to dig. Yet it has no back limbs, save for some vestigial bones hidden away, so the Mexican mole lizard spends its days squirming and paddling those front limbs through the sand. Its name may not invoke the magic of the pink fairy armadillo's, but hey, they can't all be winners.

As bizarre as it appears, the pink fairy armadillo is wonderfully adapted to subterranean life, so much so that it shuns the world aboveground. (You often hear the phrase "survival of the fittest" bandied around, which would imply that only physically strong species survive. In fact, those that are best adapted to their environment do. The armadillo isn't as much physically "fit," in the sense that it works out a lot, as it is well adapted, as we shall see.)

Accordingly, humans almost never spot the armadillo. Conservation biologist Mariella Superina, for instance, hasn't ever glimpsed one alive in the wild, and she's the *world expert* in the darn thing. Scientists can't even declare it endangered because there isn't enough data to prove it. And while you might think that a life of such obscurity—hidden away from the desert's burning days and freezing nights—would be nice and peaceful and cozy, it's anything but.

You see, when the pink fairy armadillo is burrowing, just a half foot underground, it doesn't get the relatively stabilized climate that deeper burrowers get, farther removed from the elements. When the soil heats up in the day, the armadillo heats up, and when temperatures plummet at night, the armadillo freezes. Critters aboveground seek shade for coolness or huddle together for warmth, but the pink fairy armadillo has no such luxuries. Instead, it has turned its armor into a delicate, flimsy radiator that serves as no protection from predators whatsoever.

People hate it when I tell them this, since it tends to spoil the charm of the pink fairy armadillo, but its shell is pink on account of the blood showing through. A whole lot of blood. It shows through because, like many subterranean creatures, the armadillo has no need for lots of melanin, the dark pigment that helps protect skin like ours from UV damage. All that blood is swirling around in the shell to help the armadillo thermoregulate. By virtue of being a mammal, its body maintains a certain temperature, but because sweating ain't going to do it no good nohow down in the dirt, the armadillo has evolved a different solution. If it needs to lower its temperature, it can pump blood out of its core and into the shell. This cools the blood just like a car's radiator would do for engine coolant, only the armadillo is using the soil above it as the agent instead of air. Should the armadillo need to warm up when temperatures drop at night, it can pull the blood out of the shell and add it back to its core.

Now, it seems counterintuitive for desert temperatures to fluctuate so dramatically—and rapidly, for that matter—when you'd expect the environment to be hell-hot around the clock. But it all comes down to moisture. Deserts like the one the pink fairy armadillo calls home of course have little of it, both in the land and in the air. The problem is that moist air holds heat well, which is why the climate is stable day and night in the tropics. But in the dry air of the desert, the heat quickly dissipates. That's great for the beasts that emerge only at night to hunt or graze, kicking back underground or in shrubbery by day.

While biologists believe the pink fairy armadillo to be largely nocturnal, the animal isn't about to go running around in the cool air. This is because it's essentially worthless aboveground, and I say that with all due respect. It takes moxie to live your entire life crawling through the dirt, hoovering up bugs and roots and such. Out in the big bad world, though, being ghost white isn't ideal, and having stubby legs and cumbersome claws doesn't help either. So the pink fairy armadillo probably surfaces only when it rains—according to locals who have had the honor of glimpsing the tiny rarity—in part because its burrows can flood, but also because the moisture is a detriment to thermoregulation. (The aforementioned armadillo expert Mariella Superina once filmed a nature special in which the crew sprayed the desert with jets of water, hoping that'd force the creatures out. It didn't work, but I'm sure the plants at least appreciated it.) In addition to relying on that rosy shell to maintain its body temperature, the armadillo also must keep its luxurious fur dry or risk freezing to death.

My apologies. This is getting depressing. It's no way to treat one of the most charming, most magical creatures on Earth. So I must say: Far beyond the radiator shell, the pink fairy armadillo is marvelously fit for life in the arid underworld. But if it wins for Most Adorable Subterranean Desert Mammalian on Earth (we're working on the name), there must be a second place . . . and a third place . . . and a dead last.

Our next burrower is definitely a dead last.

Naked Mole Rat

PROBLEM: The life of the burrower is plagued with tight squeezes.

SOLUTION: The naked mole rat has evolved extremely stretchy, loose skin that allows it to move better in its tunnels. Oh, also: The starch that makes this possible, hyaluronan, bestows the animal with a near immunity to cancer.

Now's as good a time as any to announce that I have an excessive amount of elbow skin. Not anywhere else, mind you—just my elbows. I can pull on it and it's, like, kinda stretchy. I'm ready to admit that.

If my elbow were an animal, it'd be the naked mole rat. Whereas the pink fairy armadillo is darling, the naked mole rat is pretty much a wrinkled sausage with way too much skin and massive buckteeth that grow outside of its lips, allowing the creature to gnaw through the dirt without choking to death. To its credit, though, the naked mole rat is not entirely bare: It has fine hairs here and there on its saggy body that function as sensory whiskers down in the blackness. Thus, it has no real use for eyes, so they've atrophied into beady little peepers. It isn't hard, then, to understand why the beast comes in dead last in the competition for Most Adorable Subterranean Desert Mammalian on Earth.

Now, when I was talking about the pink fairy armadillo I mentioned other tactics that subterranean animals use to deal with the heat and cold: burrowing deeper and huddling together. The naked mole rat does both of these. It lives in societies of as many as

THE EYES HAVE IT

It's common for critters that live in caves or burrows to have either atrophied eyes or no eyes at all. These may seem like structures evolution would want to hold on to just in case, but keep in mind that it takes a whole lot of energy and resources and time to build such things. Having the eyes evolve away frees up all this for other pursuits, plus it's one less thing to get injured or infected, which is a particularly big problem when you're rooting around in the muck. Accordingly, the blind mole rat, another burrower that isn't related to the naked mole rat, has done away with its eyes almost entirely by growing fur over them. It's a bit like wearing sunglasses all the time, only they're made of skin—and you can't take them off. All right, maybe they're nothing like sunglasses.

three hundred individuals, in a complex system of tunnels that go down to six feet deep, compared to the pink fairy armadillo's modest six-inch-deep burrows. Should things cool down, the naked mole rats can huddle together in their chambers. And should things heat up, they can retreat deeper into the burrow where there's some measure of climate control.

And indeed, the tunnels of the naked mole rat are busy places to be, with creatures shuffling around like blood cells coursing through veins. The routes can be exceedingly narrow, and this could well have driven the evolution of the burrower's floppy skin. Being so loosey-goosey allows the naked mole rat to more efficiently squeeze through its tunnels without shredding its skin—not an unwelcome adaptation if you're trying to escape predators that are penetrating your bunker.

The stretchiness is all thanks to a starch called hyaluronan. In animals it helps form what's known as the extracellular matrix—a network of sorts that holds all our cells together—and is part of the reason why I can do weird things with my elbow skin. But in a naked mole rat's skin, hyaluronan molecules are ten times longer than our own. Thus the naked mole rat's entire surface is highly stretchy. This unique solution to a problem seems to have imparted a welcome side effect on the critter: A naked mole rat will almost never get cancer.

Normal hyaluronan, in, say, us humans, communicates with cells and tells them to divide. But the elongated hyaluronan molecules in the naked mole rat *prevent* cells from dividing. And that's a cancer researcher's dream. Simply put, cancer is the abnormal division of cells, which isn't happening in the naked mole rat's body. If scientists can figure out how to supercharge the hyaluronan in our own bodies, they may be able to prevent the runaway growth of tumors. While the research is still in its infancy, one day the naked mole rat could lead science to huge breakthroughs in the fight against cancer.

Having a near immunity to cancer is great and everything, but you might be thinking that being naked is somewhat of a handicap. And you'd be right. But this is where the huddling comes in. Naked mole rats get pretty intimate down there in their burrows, amassing in congregations that sometimes grow to four individuals deep. Yeah, it sucks to be the ones down there at the bottom gasping for air, but these creatures have evolved a tolerance to low oxygen. And without the warmth of the crowd, individuals would perish when temperatures plummet at night.

That's because even though the naked mole rat is a mammal, it can't regulate its body temperature. While it would appear to be a waste to throw away the ability to maintain a constant temperature—known as homeothermy—it's not all it's cracked up to be. Sure, reptiles have to burn time sunning themselves each day to warm up, but for mammals, maintaining an internal furnace is extremely energy intensive. That fire demands fuel, which is in short supply for naked mole rats as they burrow around hoping to stumble upon a tuber to chew on. Indeed, the energy cost for burrowers to find food can be as much as four thousand times that of surface dwellers. So to cut back on energy consumption, the naked mole rat seems to have forsaken its homeothermy in favor of cuddling.

And there's no greater cuddler than the queen. Just like the pistol shrimp, these animals form a rare (for a noninsect, at least) eusocial society ruled by a matriarch, who's longer than her subjects because the spaces between her vertebrae expanded when she came into power. She's the only female that breeds, and she moves through the colony each day asserting her dominance by biting everyone else and giving them a good shake. But being both a ruler and a busy mom with embryos to worry about, she needs to

A ONE-TRACK-MIND KIND OF PERSON WHO LOVED TO BE SCRATCHED

Another much larger mammalian burrower is the wombat of Australia, a stocky teddy bear of a creature that you can't help but want to befriend. And in fact at least one naturalist has. In an account from 1963, Peter J. Nicholson describes crawling through their burrows and making the acquaintance of a young male. "Occasionally he would come up to me and sniff my arms and examine my face and hair inquisitively while I imitated his friendly grunt," he wrote. For three months the pair followed each other around, sometimes sitting at the edge of the burrow taking in the day, perhaps with an air of introspection. "He gave me the impression of being an intelligent, one-track-mind person," Nicholson concluded, adding that he "used to love to be scratched."

be sure her body temperature is ideal, so she spends more time lazing in the huddles than any other individual.

Between losing its ability to regulate its body temperature and largely losing its eyes, the naked mole rat is a reminder that there's no such thing as "progress" in evolution. (Darwin resisted using the word "evolution," from the Latin meaning "unfolding" or "unrolling," because it might imply some sort of march toward perfection. He preferred "descent with modification.") Sure, over the 3.8-billion-year history of life on Earth organisms have gotten more and more complex since that initial primordial soup of microorganisms, but intricate evolutionary innovations like the eye can fade away when they're not needed. Darwin's idea here that species don't necessarily become perfect, only well adapted to their environment like the seemingly backward naked mole rat, wasn't . . . how should I say this . . . well received. It meant that humans were just another animal that happened to evolve an impressive mind, not a creature favored by a higher power. He was right, though. We're not special, however much we think we are. We're naked like the mole rat—only we invented clothes.

CHAPTER 5

Turns Out Getting Eaten Is Bad for Survival

In Which Fish Choke Sharks to Death with Snot and Geckos Do Their Best Impression of Leaves

The statistics don't lie: Getting eaten is a leading cause of death in the animal kingdom. It's been that way for the eons that life has graced the Earth. Which is a whole lot of time for evolution to come up with some impressive solutions to the problem of predation. There's a salamander, for instance, that can regrow entire limbs that predators (or its salamander friends) have made off with. And yes, in case you were wondering, its secrets could one day get us humans to regrow our limbs as well.

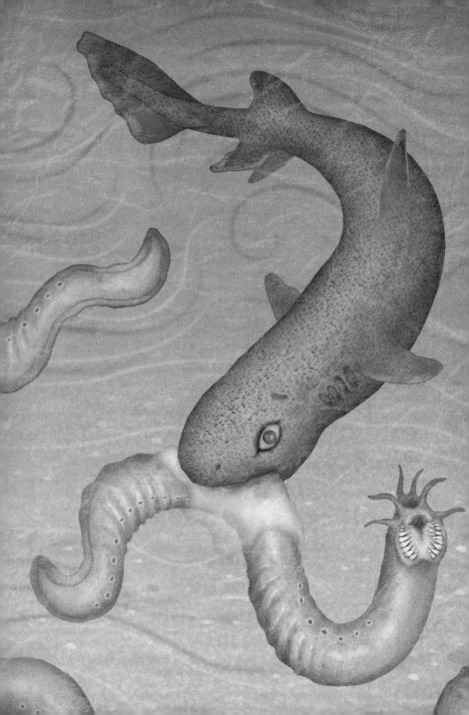

Hagfish

PROBLEM: Sharks are pretty much just giant teeth wrapped in bad attitude.

SOLUTION: The eel-like hagfish chokes its attackers to death by filling their gills with copious amounts of snot that it ejects out of glands in a fraction of a second.

Say what you will about crummy neighborhoods, but at least they're consistent. They can be extremely hot and dry or frigidly cold, you can be sure of that. But an organism can adapt to these things, for something like a desert is predictable. More befuddling, though, are the predators evolving right along with you, adapting to your adaptations, developing bigger teeth if you develop armor, or better senses if you get stealthier. And it's in response to these fiends that evolution starts getting *really* creative with its solutions. Like, evolving-to-choke-sharks-to-death-with-your-snot creative.

It got a bit shortchanged with its name, and the hagfish deserves more respect. The eel-shaped creature is a denizen of the seafloor, where it hoovers up worms and partakes in the occasional fish carcass. And when the hagfish comes upon such a bounty, the scavenger will use it for everything it's worth. It'll burrow into the body and consume the carcass from the inside out—and not just with its mouth. The hagfish's skin will absorb the nutrients given off by the rotting flesh.

But here's the problem. If the hagfish has its face buried in a carcass all day, and its eyesight is for the most part worthless, it's at the mercy of the deep's predators, including sharks and fish and eels. And quite frankly the hagfish looks pretty tasty, on account of being a wiggly tube of flesh. Luckily, though, it has one of the more novel defenses in the animal kingdom: weaponized snot.

The hagfish's sneeze is a sneeze to be reckoned with. Its body is packed with over one hundred specialized slime glands, and should you be unfortunate enough to so much as rub up against the creature, these glands work in concert to immediately spew a swirling cloud of goop. That might not seem like a big deal to you and me, say if you had the honor of handling one in a tank and playing with its slime (yes, scientists get paid to do that, and I envy them for it). But for a creature with gills, like a hungry shark, this could mean death by suffocation.

The slime is a very special one indeed, and more on that in a second, but first we should talk about how gills work. Think of them as lungs turned inside out: Just as our lungs have structures

A SLEEPING BAG OF SNOT

The parrotfish calls the more paradisiacal coral reef its home, but it ain't about to wait for an attack to unleash its own cloud of snot. When it falls asleep among the corals, it secretes its own sleeping bag in the form of an enveloping cocoon of mucus. Should a predator come sniffing around, that snotty sleeping bag could well restrain the parrotfish's scent. Plus, it doubles as an alarm system if something literally invades the fish's personal bubble. So while you may not appreciate snot, there are at least two creatures on this planet that certainly do.

that are packed with capillaries, where the blood pulls in oxygen from the air, fish gills are packed with capillaries that make direct contact with the water. We humans have to keep air flowing through our lungs at all times and fish have to do the same with their gills, so they can't just lie there and expect to gather enough oxygen. This is why you'll see fish gulping—that's them pushing water over the capillaries. (Sharks have to swim around to keep the flow going, but it's a myth that they'll all drop dead if they stop swimming. Open-ocean species need to, sure, but other varieties, ones that rest on the seafloor, will instead gulp air like their bony-fish cousins.)

So, the slime. Each of those hundred-plus slime glands is packed with two different kinds of cells. One type produces your run-of-the-mill mucus, but the other ejects incredibly strong threads that are each six inches long. I'll reiterate. A microscopic cell that measures 0.004 inches long by 0.002 inches wide fires out a thread a half foot long. That's . . . infuriatingly illogical.

But it works. The secret is how the thread is packed, spun precisely around and around on itself to form an oblong yarn ball held together by a water-soluble glue. When the hagfish is in trouble, it contracts the muscles around the slime glands, ejecting the mucus and the thread bundles, which unravel as the glue dissolves in the water. The released energy from the bundles helps inflate the cloud, which only gets bigger as the hagfish flops around trying to escape.

All told, there are some six thousand threads per cup of slime, and these strands easily tangle in, say, the gills of a predator. The defense works so well that scientists originally began to figure the goo served this purpose—as opposed to the hagfish releasing a

THAT TIME A DENTAL DAM LED TO A SCIENTIFIC DISCOVERY

One experiment that confirmed hagfish slime to be a defensive weapon is probably the most amazing study I've ever read, at least as far as materials are concerned. It involved a rockfish's disembodied head, a hagfish, PVC pipe, a siphon, and an "extra-heavy dental dam." Researchers wanted to see if the hagfish's slime really had an effect on water flow over a fish's gills. So they covered the end of the pipe with the dental dam, rammed the fish's head through, propped its mouth open, added a siphon at the other end, and dropped the whole weird apparatus into a tank with a hagfish. The scientists switched the siphon on and our brave hagfish "was pinched on the tail with padded forceps to induce sliming." The siphon sucked the resulting goo through the fish's mouth and into the gills. The rockfish gave up its dignity and the hagfish its slime—not to mention the invaluable conclusion that its kind had in fact evolved to weaponize snot.

cloud and hiding in it like it was a force field—because they wouldn't find the critters in the stomachs of fish, but instead in those of gill-less creatures like dolphins and seals. Another bit of good evidence is that the hagfish will only fire a mucous gland that something has agitated. Squeeze the creature and goo comes out at the points of contact, not anywhere else. Substitute your fingers for the jaws of a shark, and you have a well-targeted countermeasure against predation.

Seeing video of unfortunate sharks and fish tangling with the hagfish really makes you feel for the would-be predators. Their reaction comes almost instantaneously after the bite. The swirling cloud explodes, and there's much head shaking and gaping and full-body convulsions and a rapid retreat, with the attacker trying desperately to dislodge the snot from its face. And if it fails, it'll suffocate to death and sink to the seafloor to become a welcome buffet. The isopods and crabs and bacteria will come from all around—joined by, inevitably, the lowly hagfish.

Axolotl

PROBLEM: A leg is a terrible thing to waste.

SOLUTION: The axolotl salamander can regenerate entire limbs lost to predators.

Lucky for me, the salamander is already dead. And lucky for the salamander, really, because as it would turn out, I'm not that great at amphibian surgery. But I take a deep breath as a sharpshooter might, pull the limb nice and taut, and with scissors cut through it. Well, mostly through it. I'm looking through a microscope, so my perspective is all off. The scissors get through the humerus all right, the bone not so much snapping or cracking like I'd expected, instead just gently crunching. But the tips of the scissors meet before they're all the way through the limb. In a panic, I pull the limb taut again and make another cut. But still unaccustomed to the magnified view, I miss again. So I cut again. And reorient again and miss again. And snip once more to finally liberate the limb.

I pull my eyes back from the microscope and look down at the four-inch-long salamander, all clad in white tissue paper save for the deep-red external gills erupting out of the back of its head and the bloodied bits around the stub. It's an axolotl of Mexico. Cut off one of its limbs, and in a month it'll grow right back, every bit as functional. Remove part of its brain or jaw or spine, and that'll all grow right back, too. The axolotl can't be bothered to not remain in one piece.

I should clarify here. I didn't just desecrate an axolotl in some back-alley pet shop. I'm in the Limb Regeneration Lab at the Uni-

THAT TIME THROWING A SALAMANDER IN A FIRE LED TO . . . ABSOLUTELY NOTHING

The salamander has long been revered in European folklore not for its powers of regeneration, but for its supposed immunity to the ill effects of fire, e.g., burning to death. Legend said it was born from flames, which is metal as hell but probably not true, considering that amphibians—with their moist, delicate skin—are among the animals most sensitive to fire. There was even a naturalist who put this all to the test: the great Roman writer Pliny the Elder. And by "test" I mean he tossed a salamander in a fire, with predictable results. Sadly, the path to human knowledge seems to be paved with things like charred salamanders.

versity of California, Irvine, funded by the one and only US Department of Defense, which longs to bestow the powers of regeneration upon wounded soldiers. These are brilliant scientists who let me go to town on a dead specimen, as I am clearly unqualified to perform surgery on a live salamander. And these scientists don't think it's a matter of if humans can grow back limbs like the axolotl can, but when they figure out how to do so, for we have the exact same genes that give the creature these powers. And the scientists believe that these procedures on salamanders (the living variety, not the one I desecrated) will get us there.

Here's the thing about evolution that never ceases to blow my mind. In a way we're related to every single organism on Earth, be it an animal or plant or bacterium, because the tree of life began with one organism and over billions of years branched out into the incredible array of creatures we know today. Over the millennia, as creatures evolved and new species split off and went down their own lineages, they still shared what's known as a common ancestor with the animals they diverged from. So we share damn near 100 percent of our DNA with chimpanzees because we have a relatively recent common ancestor, but we're more genetically distinct from the axolotl, with which we share a much more ancient common ancestor. Yet still we have the genes required for regeneration. In a way we're part salamander—we just don't use those genes the way the axolotl does.

The benefits such regeneration bestows on the axolotl are obvious. The salamander is an aquatic species, with plenty of predators to worry about. (Well, it *did* worry about them at one point. The axolotl may well exist only in labs now, the expansion of Mexico City having forced it out of its lake habitats.) If a fish makes off with a limb, growing the thing back allows the salamander to keep clambering around the lake floor. And that's not the axolotl's only worry: They tend to eat each other's limbs. Like, really frequently. So growing back amputated limbs helps the axolotl continue hunting in order to survive, not to mention partaking in a complicated mating ritual.

Sounds good, right? Let's get us some regeneration, since, after all, we have the same genes? Well, it isn't as easy as all that.

Instead of regenerating, we humans scar, and that's suited us perfectly well over the course of our evolution. But the axolotl never, ever scars. Well, I'll qualify that: Its regeneration is a series of steps, the first of which is a bit of scarring, but that disappears as the process unfolds. But why? Why do our bodies get so carried away with scarring, when the axolotl shows that it's possible to switch at some point to regeneration? The science here is still murky, but a clue could be that when the axolotl is regenerating, its immune system takes a hit. The salamander is coated in a powerful antibiotic mucus to compensate for that, but we humans don't have such a luxury. Growing back a limb won't do you any good if you die of an infection first.

WHEN 894A MET 664A:
AN AMPHIBIAN LOVE STORY

It was my great honor to witness the axolotl mating ritual in that university lab, in a tale I like to call "When 894a Met 664a: An Amphibian Love Story." A technician has placed specimens 894a (to be known henceforth as "the male") and 664a (to be known henceforth as "the female") in a tank together with some nice rocks and plastic plants for decor.

At first the lovemaking is slow going. Neither moves much, instead mutely staring and flicking their fluffy gills back and forth. But then it happens: the first contact, as the male ever so slowly pulls his head closer to her, nudging her side. He then makes for her cloaca, jamming his snout under her body and nuzzling her naughty bits, nearly lifting her onto her side. But seemingly frustrated by the female's indifference, the male freezes, then violently flicks his tail to rocket away. He hangs out alone at the opposite side of the tank for a while. More stares between the two abound, until the male again approaches the female, turns to look at me as if to ask for help or at the very least some advice, for Christ's sake, and vomits. Love story over: 894a and 664a, it would seem, were never meant to be.

What's clear about axolotl regeneration is the importance of what are known as growth factors, proteins that promote the development of tissue. Take your tweezers and scissors and amputate an axolotl's limb, and these growth factors recruit cells around the wound site to begin rebuilding the structure. It's important to keep in mind that all manner of different materials are involved here—bone, muscle, skin, etc.—and the body needs thorough instructions to get the cells into the right spots. The salamander isn't building out a uniform stick of flesh. It has to know when to start forming a joint and all that jazz.

Inducing such regeneration in humans isn't about tweaking the genes we share with the axolotl. That's complicating things. Instead, human regeneration will likely be about harnessing the power of the growth factors associated with those genes. So labs like this one are trying to decode the steps the axolotl takes to regrow a limb, so we might someday apply the growth factors to wounds in the right order.

It sounds like lunacy, I know. But again, these researchers think it's only a matter of time before the axolotl helps unlock the secrets of human regeneration. And when we do, we can all look back on that day when I wasted these scientists' time and held back the advance that much longer.

My apologies, humanity.

Cuttlefish

PROBLEM: For sea creatures, sometimes a shell just isn't good enough for protection.

SOLUTION: Cuttlefish have evolved the animal kingdom's most incredible active camouflage to imitate any kind of background in a flash.

When it comes to keeping from being eaten, critters have a couple of options. The hagfish has an active defense, opting to blow the nose that is its entire body, and the axolotl goes about things more passively, resigning itself to the occasional dismemberment. But the cuttlefish has a far more highfalutin strategy to survive predation: It avoids detection by deploying some of the most astounding camouflage on Earth.

Let's pump the brakes a bit and back up some 500 million years to meet the ancestors of the cuttlefish and other cephalopods, like the octopus and squid. Far from sporting the floppy cephalopod bodies we see today, these ancient creatures protected themselves with strong shelled armor. There was a downside to that protection, though—it made them slow and ungainly, as the only remaining shelled cephalopod, the nautilus, demonstrates by awkwardly swimming around and bumping into reefs and things like it's on horse tranquilizers.

So it could have been that a powerful predator evolved in the seas, a beast capable of crushing shells, and the sluggish cephalopods found themselves losing their advantage. These creatures, though, were no pushovers. They found other ways of surviving.

They lost their shells and diversified into all manner of species, utilizing a wide variety of strategies (the cuttlefish's ancestry is betrayed by its internal surfboard-shaped cuttlebone, which helps it regulate its buoyancy). Squids have their speed, and octopuses can squeeze themselves into almost impossibly tight crevices, but among cephalopods, the cuttlefish puts on the most audacious defense.

The cuttlefish never met a surface it couldn't blend in with: algae, coral, sand—even an artificial pattern like checkers. The transformation is almost instantaneous, and it's mind-blowing. Chase a cuttlefish into a field of seaweed and it'll hunker down, modify the color and pattern and even texture of its skin, and lie still, swaying in tandem with the vegetation. There it'll wait you out, but if you close in further still, it'll give up the ruse and rocket away, leaving a cloud of ink in its wake.

LOVING WOULD BE EASY IF YOUR COLORS WERE LIKE MY DREAM

The cuttlefish is sometimes referred to as the "chameleon of the sea," but that's a bit of a misnomer, not because chameleons can't swim too good, but because the two creatures have different uses for their camouflage: Chameleons don't just change color to blend in with their environment. A good indicator of this is the fact that chameleons can be splotched with bright reds and blues and greens simultaneously, which wouldn't do them much good as camouflage unless they're sitting in a box of crayons. Instead, they're using it to signal to potential mates and help with thermoregulation. If a chameleon is a bit chilly, it can darken its skin to absorb more of the sun's energy, and if it gets too hot, it can dial down the color again. Think of it as having a good coat that never goes out of season.

The spectacle comes from three special layers in the cuttlefish's skin. The bottommost layer is plain old white flesh. The next is a sort of iridescent surface that reflects light to provide green and blue. But the top layer is where things get interesting. It holds cells called chromatophores, each being a sac of a particular pigment—orange, red, yellow, brown, or black—to which tiny muscles are attached. When the cuttlefish wants to express a certain color, it'll contract the muscles around those cells, opening up the surface area of these chromatophores as much as 500 percent to expose more pigment.

Because these cells are hooked up to muscles and nerves communicating with the brain, the cuttlefish is bestowed with camouflage so quick it doesn't seem possible. The light show is so well orchestrated that some species can send pulsing waves up and down their bodies, perhaps to serve as a warning or, and I'm being totally serious here, hypnotize their prey. Divers have seen them flaring out their arms and deploying a rapid, chaotic light show as they approach a soon-to-be victim, then fire tentacles out that snatch the prey and reel it in.

Coordinating all of this is one magnificent brain. It should be noted that invertebrates typically aren't the sharpest knives in the drawer compared to vertebrates like ourselves. And that's okay, because they've got other things going for them, things like rugged exoskeletons and the ability to dry out and reanimate like the water bear. But cephalopods are a dramatic exception to the doltishness. They're scary smart. In the lab, cuttlefish can learn to solve mazes, while it's an all-too-common occurrence that octopuses figure out how to escape their tanks and wander down the hall. (All right, maybe not too bright as far as long-term goals go, but still . . .) Such smarts are indispensable when hunting, but the cuttlefish's formidable brain also has to coordinate its defense, processing its surroundings and translating that into camouflage.

They're such accomplished camouflagers that they can even outwit members of their own species. The male giant Australian cuttlefish, for instance, cross-dresses to get laid. Big males hold sway over the females, guarding them and attacking any rivals that

IF IT HAS "COLOSSAL" IN ITS NAME, IT'S PROBABLY BAD NEWS

I don't want to sound like an alarmist, but you should never go in the ocean ever again, for the cuttlefish has enormous cousins that will eat you: the colossal and giant squids. Okay, fine, there's never been an attack on a human, but these squids can balloon to sizes that defy belief. The giant squid reaches forty feet long (though, in fairness, a good amount of that is just two long tentacles), and although the colossal squid grows to just fourteen feet, it's far bulkier than the giant, weighing in at over one thousand pounds, making it the heaviest invertebrate on Earth. The giant squid has suckers with serrated edges (their mortal enemies, the sperm whales, often have ring-shaped scars around their mouths), but some of the colossal squid's suckers sport something far more sinister—swiveling hooks that sink into flesh. And here you were thinking I was an alarmist.

approach. But smaller males will modify their color to mimic females, and also morph their arms in such a way as to appear to be holding an egg like the ladies do when they're not keen on mating. The sneaky male thus hits a middle ground: The dominant male won't attack him because he thinks the imposter is a female, but he won't try to mate with him because he thinks he's an *unreceptive* female. So the cross-dresser slips under the big male and mates with his female, passing along his genes for brains, not brawn.

Oh, I almost forgot: Cuttlefish are also color-blind. I'll let that sink in, then disappoint you by saying that no one is quite sure how that's possible when the creature is perfectly matching its color to its surroundings. A 2015 discovery regarding the California two-spot octopus, though, could provide a clue. Researchers removed patches of the octopus's skin—which is also packed with chromatophores—and exposed them to light in the lab. The chromatophores expanded on their own, obviously without the help of the creature's eyes or brain. Furthermore, the scientists found that the octopus's skin is loaded with the same light-sensitive protein you'd find in eyeballs, so it would appear that somehow the skin itself is processing color. Whether the ability also extends to cuttlefish is to be determined, but I'm betting it's likely.

The cuttlefish may not be seeing color with its eyes, but what it is most certainly picking up are contrasts in the environment. Curiously, for all its transformations, the cuttlefish is working with only three pattern templates of varying contrasts. There's "uniform," which is pretty much a single color; "mottle," which looks like static on a TV (remember the days of static on TVs?); and lastly, "disruptive," which splits the skin into a checkered pattern. Using these templates and adding in the appropriate colors, the cuttlefish can dissolve into whatever background it sees fit, deploying the uniform pattern for something like monochromatic sand, the mottle for more multicolored sand, and the disruptive for more complex backgrounds like coral.

Thus the cuttlefish, bereft of a protective shell, manages to stay out of stomachs. Some other camouflagers, though, aren't going through the trouble of deploying such light shows. They pick an outfit and stick with it, and there's no better-dressed camouflager than the satanic leaf-tailed gecko. Yes, that's what it's actually called. And yes, it's every bit as amazing as its name would imply.

Satanic Leaf-Tailed Gecko

PROBLEM: Like the tenrec, geckos in the forests of Madagascar face predators aplenty.

SOLUTION: The satanic leaf-tailed gecko does a startlingly faithful impression of a leaf, complete with veins and all.

There's a picture floating around the Internet of what appears to be a tiny dragon holding on to a branch, with wings sprouting from its back and skin that alternates between red and black. Its eyes, too, are a demonic red, and its tail has chunks missing as if burned away. The photo is, as you can imagine, a fake—but not as much as you'd think. The hoaxer just added the wings and maybe took a few liberties with the saturation. The rest of the creature is perfectly real.

The photo is of a satanic leaf-tailed gecko, one of fourteen known species of leaf-tailed gecko that call Madagascar home. For my money, these critters are some of evolution's greatest triumphs, so well camouflaged that it's scary. Not scary because they can tend to look like they'd eat your soul, but because they demonstrate what a powerful force natural selection is.

I doubt you'd be able to tell a satanic leaf-tailed gecko from a cluster of leaves if you weren't looking for it. The ridge of its back is white, from which radiate white lines—the veins of the leaf. Its tail is shaped like its own leaf, frayed at the edges, seemingly rotted away. Twisting its body in on itself, the gecko completes the ruse: If it didn't have flesh and blood and organs and all that, it'd be an honest-to-goodness dead, curled-up leaf. But how could such camo evolve? Did the gecko have to, like, think about it really hard? It didn't, and it wasn't placed on Earth as a gag. The satanic leaf-tailed gecko is an absolute marvel of evolution, and it's all thanks to sex.

I failed to mention a while back when we were talking about creatures getting laid that there are a lot of animals that don't get laid at all. They're asexual, opting to clone themselves instead of finding a partner. For instance, a gelatinous little aquatic creature called a hydra—which looks like a palm tree, only with fronds that are stinging tentacles like a jellyfish's—will bud little versions of itself that break free and go about their own lives. Asexuality is fantastic if you aren't able to find a mate to fertilize you, but it does have a very serious downside.

ON THE ORIGIN OF SPECIES

So we've been talking about all of these species, and tried and failed to define what a species is (not my fault), but how do we get a species in the first place? Well, typically it's a matter of isolation. Take the satanic leaf-tailed gecko and the many weird creatures of Madagascar. When the island drifted off, its species found themselves isolated, and in their isolation they diverged from the populations they left behind. Over time, they became genetically distinct enough that they were no longer able to breed with their counterparts. They'd speciated. This can happen on the mainland as well when, say, a river cuts through a habitat or a mountain range grows, splitting a population in two. And voila, a shiny new species.

Beyond feeling good, sex is great because it kicks evolution into overdrive. A hydra's offspring are clones—that is, they're genetically identical to their parent—but the offspring of sexual reproducers get a random mixing of genes, which introduces a bit of welcome chaos. You're beautiful and your siblings aren't so much (or vice versa, heaven forbid) in part because when you were conceived, your parents' genes came together in random ways. Such variability between offspring helps ensure that at least a few of the kids have the traits needed to survive in a given environment (not that I'm saying that being good-looking helps humans survive, though it'd certainly help you pass along your genes).

So, for a satanic leaf-tailed gecko in the rough-and-tumble jungles of Madagascar, not looking exactly like its brothers and sisters can make all the difference. The species has set down the evolutionary path to mimic a leaf—not deliberately, of course—so even the most minute variations count. A vein here or a chunk of a tail missing there, a slightly better color to match a dead leaf—every little detail is nothing to sneeze at when the geckos have hungry eyes locked on them. These traits help the geckos survive, so through generation after generation after generation, individuals with subpar camo get knocked off, while those with beneficial variations pass down their genes.

Thus something as perfectly camouflaged as the satanic leaf-tailed gecko can appear to be designed. But it's just a matter of predation pressures. While these are nocturnal hunters, relying on their giant eyes to gather what scant light is available in the rain forest, they still have to be able to make it through the day. So, depending on the species, a leaf-tailed gecko will make use of whatever camouflage evolution has bestowed on it to disappear when the sun comes up, ideally slumbering unperturbed. The more leaf-like species curl up among dead leaves, while others lie flat on tree trunks to blend in with the bark, and still another—the mossy leaf-tailed gecko, with its speckled skin and frilly edges—seeks out moss-coated bark. None can modify its camouflage on the fly like the cuttlefish can, but the disguise is so convincing that it doesn't matter.

Similar-looking species of leaf-tailed gecko roam Australia, but geckos around the world for the most part are comparatively plain Janes, sporting green or brown or yellow skin, sometimes

SGT. PEPPER'S LONELY LIGHT-COLORED MOTHS CLUB BAND

You might think creating something as complex as the satanic leaf-tailed gecko would be a drawn-out process that takes thousands or millions of years, and you'd be right. Evolution can take its sweet time with things. But then again, it can also progress with incredible speed.

Take the peppered moth of Britain, a speckled creature adapted to hang tight on lichen-covered trees, thus avoiding birds. But it wasn't expecting the Industrial Revolution and the consequent pollution, which coated trees in soot. The peppered moth, though, would not fade quietly into extinction. It adapted, developing a darker coloration that was first observed in 1848. Just fifty years later, 98 percent of peppered moths were dark. And when clean-air laws took effect in the twentieth century, the moths went back to being speckled.

The whole saga was evolution gone turbo. Predators were more likely to spot light moths when the environment became coated in soot; thus, the moths' darker counterparts prevailed and gave rise to soot-colored babies. When the habitat was clean again, dark moths struggled to survive; thus, light moths prevailed. And that, ladies and gentlemen, is an exceedingly rare pollution success story.

with stripes, other times with polka dots. Yet nothing, not even the Australian varieties, can touch the wonder of the satanic leaf-tailed gecko. But you'd think that if it's possible to evolve such an outfit, it'd suit geckos around the world to try it on. So what's so special about the satanic leaf-tailed gecko?

No one knows for sure. But it's worth noting that Madagascar is a very strange place with all manner of goofy creatures found nowhere else on Earth (accordingly, this isn't the last you'll be hearing of the island in this book). Islands are fascinating like that. When Madagascar went independent 90 million years ago, its ecosystem reset. The species that rode it out to sea and managed to cope with the isolation evolved uniquely in the new order. And other species, like birds that could manage the expanding oceanic gap, joined them from time to time. Species came, and species went. In their isolation, these creatures evolved like nowhere else on the planet. So maybe there's something about the predators in Madagascar—maybe they're that much more menacing—that made the satanic leaf-tailed gecko get so carried away with its camouflage. Or maybe, just maybe, a higher power got drunk and decided to hide some geckos in the forest.

Seems unlikely, though.

Pangolin

PROBLEM: Ever heard of something called a lion?

SOLUTION: A mammal known as the pangolin has developed gnarly-looking keratinized armor that's impervious to lions, not to mention the ants it eats.

Jump a few hundred miles from the island of Madagascar across the Mozambique Channel and if you're lucky you'll find an African critter that's stumbled upon a different solution to the problem of predation. The pangolin is a mammal, but you'd be forgiven for mistaking it for a lizard. It's a streamlined critter covered head to toe with overlapping scales that give it the look of a walking roof, or maybe an artichoke. The thing has shingles—actual shingles, not the rash.

The pangolin is a living tank. When threatened, it will curl up into a tight, almost spherical ball, wrapping its long armored tail around itself and waiting out the attack. The defense is so effective that even a lion, one of the most powerful predators on Earth, can't break through. It'll bat at a pangolin a bit and give it a gnaw here and there, but the lion does so at its own peril: Those scales are crazy sharp and unfriendly to mouths. Sometimes the rest of the lion's pride will join in, taking turns trying to pry the pangolin open. But it's all for naught. Inevitably they give up and trot off, and the pangolin uncurls itself and goes about its day.

CHARLES DARWIN DISCOVERS THE VW BUG

Contrary to popular belief, Charles Darwin was not the official naturalist when he boarded the *Beagle*. That honor went to the ship's doctor. In reality, he seemed more enthralled with geology than biology during the journey, devoting a great deal of his account, *The Voyage of the Beagle*, to rocks. His worlds of geology and biology collided, though, when he happened upon the fossilized remains of an enormous armored mammal: *Glyptodon*. It would have been about as big and heavy as a VW Bug (not Darwin's comparison, obviously), "with an osseous coat in compartments, very like that of an armadillo." Unlike the pangolin with its many overlapping scales, *Glyptodon* had a solid shell of armor, like a mammalian tortoise. It was in fact a relative of modern armadillos and played no small part in shaping Darwin's theory of evolution by natural selection. He realized a higher power didn't place species on Earth and wipe them out for the hell of it—there's a great chain of relatedness among creatures.

■ THE WASP THAT BRAINWASHED THE CATERPILLAR

Even humans and our many tools have a hard time breaking their defenses. The American naturalist William Beebe (our friend who had a few words to say about the anglerfish) once encountered one in Borneo, noting that the curled-up pangolin's tail muscles "were as rigid as steel." William's guide tried to open up the creature with a shovel, but "even with the spade as leverage, hardly an inch could be pried free." They left the poor thing alone for five minutes, and then it suddenly raised up its scales, which any self-respecting naturalist would be irresponsible *not* to tug at. Unfortunately for William, "like the jaws of a steel trap they closed down with such force as to bruise the finger cruelly, or actually to pinch off a fragment of flesh." That seemed to tickle his guide: "One's enthusiasm for scientific investigation in this direction was satiated at one trial, which was also sufficient to prove that a Tamil trailmender has a sense of humor and a lack of sympathy."

These scales are made out of keratin, a wonderfully useful protein in the animal kingdom. Creatures have incorporated the stuff into all kinds of things, from our own hair and nails to a deer's hooves to the horn of a rhino. It makes up a bird's beak and the filter-feeding baleen structures in some whales. Keratin is *everywhere*, and it's both strong and pliable, as you might notice with your nails. So instead of snapping in half when a lion's claw (itself made of keratin) snags it, the pangolin's scales flex, absorbing the energy. Indeed, its strength has not been lost on our species: In India, where the pangolin also roams, warriors once wore armor made of the creature's scales.

The pangolin's formidable coat serves another more cryptic purpose. The creature is an insectivore, specifically an insectivore that targets eusocial insects like ants and termites. Using its huge claws, it tears into colonies and laps up the swarming insects with a long, tubular tongue. The thing can be as long as the pangolin's body minus the tail, but contrary to what you'll read pretty much everywhere on the Internet, it does *not* attach near the animal's pelvis, which, granted, would be kind of awesome. Instead, the tongue is attached near the bottom of the sternum, and it's coated in mucus from two massive salivary glands. Firing the tongue into ant or termite colonies, the pangolin can get into tunnel after tunnel, snagging the infuriated, panicked insects and reeling them in. The bugs swarm and try to sting and bite but find no purchase on the pangolin, whose scales protect it from the onslaught. The pangolin's eyelids are also nice and thick to keep out stings and bites to the eyeball. Its ears, too, have special valves that seal to keep out attackers.

But the brutal irony of the pangolin is that the scales that help it survive could well drive it to extinction, all thanks to—drumroll, please—humans. In traditional Chinese medicine, pangolin scales are prized for a wide range of "cures." Is your kid being difficult? Get some pangolin scales. Are you deaf? Pangolin scales. Have you been possessed by demons? Pangolin scales. Suffering from malaria? Yep, pangolin scales. Just roast them, grind them

HAVE YOU SEEN MY WIFE AROUND? SHE'S GOT SCALES AND LOVES ANTS

The pangolin holds a special significance among the Shona people of Zimbabwe: Legend goes that long ago, chiefs sent the wife of a spiritual medium, known as a midzimu, to roam the bush, where she became a pangolin. So if you stumble upon such a creature, you're required, at risk of bad things happening to punish you, to grab it and take it to the midzimu, since it could well be the very same wife. That's why pangolins curl up into a ball when approached—instead of fleeing, the wife is signaling that she wants her discoverer to pick her up and take her back to her husband. The midzimu then proceeds to cook and eat the pangolin. Not sure how that makes sense, but hey, who am I to judge?

up, and mix the ash with oil or butter or even a boy's urine (presumably not that of the bratty kid you're trying to cure—that would be counterintuitive).

Demand for its scales has made the pangolin the most trafficked mammal on Earth, with all eight species considered endangered or critically endangered. As if the market for the scales weren't enough, pangolin meat is all the rage in Asia. China's booming economy in particular has brought with it huge demand. Wild meat like that of the pangolin is a luxury, eaten to celebrate business deals.

Unless something is done soon, the world's toughest, most wonderfully armored mammal will slip into oblivion. William Beebe saw it coming all along, though interestingly he noted that "their flesh is too infiltrated with formic acid to be palatable." Maybe William had a particular sense of taste. Regardless, he notes that "until the excessive increase of human dominion and the consequent decrease of anthills comes to pass, the race of Pangolins will continue to flourish on the earth." We can be confident that the anthills will remain, but whether humans can get their act together certainly isn't a given.

Crested Rat

PROBLEM: Not all African mammals have the luxury of armor.

SOLUTION: The crested rat deploys special hairs that it slathers with chewed-up poisonous bark. That'll leave a bad taste in an assaulter's mouth, and could very well kill it.

There's a plant that we may as well call the Grim Reaper of East Africa: *Acokanthera schimperi*. According to one guide to African ethnobotany, it's also known as the "national poison plant" of Kenya, because anybody who "hunts with a poisoned arrow, poisoned spear, or poisoned trap and uses poisoned weapons against enemies, reaches solely for *Acokanthera*." The methods for its preparation differ from community to community, but the end result is always the same: death, and a quick one at that. The poison can even bring down elephants, and indeed certain African peoples have used six-foot bows to rocket three-and-a-half-foot, poison-slathered arrows into the beasts, dropping them dead of heart failure.

You'd think, then, that every creature in Africa would go out of its way to avoid the *Acokanthera* plant, but not a rodent called the crested rat. The beast gnaws on the bark and roots of the thing, gives it a good chew, then applies the poisoned spittle to specialized hairs on its flanks. All it takes is one bite from a predator to get the toxin into a mucous membrane, where it goes to work shutting down the assaulter's heart.

The crested rat is a perplexingly arranged animal, with fluffy fur and a hairy tail, unlike your average sewer rat. The hair along its flanks is black and white, giving it the appearance of an animal called a zorilla, which looks a lot like a skunk and even comes equipped with that characteristic stinky spray defense. Scientists have suggested that the crested rat evolved to mimic the zorilla, taking advantage of its bad reputation without having to develop a noxious spray. Problem is, the rat only exposes that black and white flank hair when threatened—otherwise the coat is more of a grayish color. So this probably isn't a matter of mimicry.

No, the crested rat has come up with its own unique defense. To expose that black and white hair, it uses special muscles to both erect the hair up and deflect it down its flanks, parting it so the rat can apply the poison. While these flank hairs may not be hard and sharp like a porcupine's or hedgehog's, they're special in their own way. They're perforated cylinders that hold many strands, which act like sponges to absorb the creature's tainted saliva. When the rat is done with the application, it lets the normal hair fall back down, enclosing the specialized black and white hairs, presumably to keep rain from washing out the poison.

Unlike the zorilla or the skunk—whose piebald colorations serve to warn predators that they aren't to be trifled with— by exposing those black and white

WEAPONS OF MASS CONSUMPTION

The crested rat may appropriate the toxins of a plant, but there's a beautiful sea slug out there, the blue dragon nudibranch, that appropriates the weapons of one of humanity's most feared oceanic menaces: the Portuguese man-of-war (which, fun fact, isn't a jellyfish but a siphonophore, a kind of colonial organism made up of a bunch of clones). But the blue dragon doesn't take the man-of-war's stinging tentacles and apply them to its body. No, it straight up eats the things. Somehow, someway, the stinging cells make their way through the walls of the nudibranch's digestive system and end up in its skin, ready to torment anything foolish enough to touch the creature. Sure, it doesn't breathe fire like a dragon, a challenge when you live underwater, but its burn is just as devastating.

hairs when threatened with attack and turning one of its sides to face its foe, the crested rat may well be *inviting* a bite. I mean, clearly it'd be ideal not to be assaulted in the first place, but if the rat is going to get bitten anywhere, it wants to get bitten on those toxic hairs. That's how confident it is in the speed with which the poison sets in.

Should the diversion fail, though, the crested rat comes equipped with a few other handy adaptations. While it may look cute and fluffy, it's built like a linebacker, with an armored skull and beefed-up vertebrae, not to mention tough, dense skin. That skin is so tough, in fact, that researchers once examined the corpse of a crested rat and found that it was heavily bruised as if a dog had mauled it, yet the skin hadn't been punctured. This is a creature that has evolved to take a beating.

Dogs seem to have a particular fondness for molesting the rats, while lacking a particular immunity to the poison. Accounts of attacks have been numerous, and often brutal. The scientists who first studied the structure of the crested rat's special hairs noted that symptoms among canine attackers "range from mild lack of coordination, mouth-frothing and signs of general distress to collapse and rapid death, apparently from heart failure." Other dogs have fared better, recovering from their stupor after a few agonizing weeks. And at least one prudent pup that "survived a putative near-death encounter displayed every sign of fearful aversion" when it came across another crested rat.

The *Acokanthera* is clearly a tree of staggering toxicity, which should therefore elicit the question: How is the crested rat itself not dropping dead after getting a mouthful of the stuff, when all that's needed for any other creature to drop dead is, well, getting a

---

HAIR-RAISING EXPERIENCES

You know how cats do that thing where they bristle their hair when they're all worked up? They're trying to make themselves look bigger to fend off an enemy. It's a brilliant little strategy, and it turns out that we have remnants of it in ourselves: goose bumps. Humans descended from some sort of mammal that raised its hair to look bigger, and when our adrenaline kicks in for whatever reason, the thin hair we have left still tries to go erect. And when you get goose bumps from the cold? Our ancestors probably used the same mechanism to raise their hair, thus trapping a layer of air to better insulate themselves. As for the geese, well, they just have bumpy skin. What are ya gonna do.

mouthful of the stuff? No one knows for sure, but a clue could be the rat's outsized salivary glands. It may be that the crested rat's saliva has some kind of protein that binds to ouabain, the compound that gives the *Acokanthera* its toxicity. We'd also expect the rat to be ingesting at least a little bit of the stuff, so perhaps its digestive system is also adapted to handle ouabain.

Getting eaten is bad, and the crested rat has stumbled upon a truly inventive manner of avoiding it. But evolution abhors an advantage. As creatures evolve defenses, their predators evolve workarounds, and as the prey evolve even better defenses, the tug-of-war goes on and on. We all must eat, after all. And when it comes to eating, our next batch of critters simply won't suffer starvation.

These are Earth's most insatiable appetites.

CHAPTER 6

It Turns Out Not Eating Is Also Bad for Survival

In Which Primates Give Us the Finger and Beetles Run So Fast They Go Blind

There is, of course, the other side to all this predation: the gluttons themselves. I mean, a foot-long snail's gotta eat, even if it means laying waste to Miami. As does a certain crustacean (not the Miami bit—the having-to-eat bit), so it's evolved to blast through clam shells with shockingly powerful punches. Think Mike Tyson, only with more legs and fewer face tattoos.

Giant African Land Snail

PROBLEM: A snail has to eat not only for energy, but to build its shell.

SOLUTION: The giant African land snail lays waste to vast swaths of vegetation. And when the snail invades Florida and can't get enough calcium to build that shell, it lays waste to the stucco in vast swaths of homes.

I want to be clear about something: I've got nothing against Florida. I went there once, and the weather was great. But Florida is a . . . singular place. Again, nothing against it, and I'm not just saying that to avoid alienating a market for this book. But whenever you hear that a guy flashed his bum at an IHOP in a quixotic attempt to get free food—and that was only after impersonating a cop didn't do the trick—you think, "Oh, Florida, it has to be." And sure enough it is (specifically, in Orlando in January 2014). Or when a guy dials 911 to check up on that tax return of his. Yep, Florida again (a month later in St. Petersburg). Or when foot-long creatures called giant African land snails maraud neighborhoods devouring houses. It could only be Florida, and I say that with all due affection.

I'VE ASSAULTED SNAILS WITH SALT, AND FOR THAT I'M SORRY

If you were once an adolescent idiot like me, you poured salt on snails and said mockingly, "I'm *meeelting*!" because you were cruel and too young to realize that's a dated cultural reference. It was terrible behavior, and I'm sorry for doing it. In reality, salt doesn't "melt" the snail. That foaming is the salt drawing the water out of the animal, which dies of dehydration. Again, I'm sorry, both for salting snails and for the dated cultural reference.

Florida has a very, very bad and very, very giant snail problem, where the critter has become an invasive fiend. Part of the problem is that these snails have a pretty good thing going. Like the penis-fencing flatworms, they're hermaphrodites, so one sexually mature giant snail is always capable of mating with another sexually mature giant snail. And like the flatworms, two snails can fertilize one another. They'll then go off and lay up to four hundred eggs, and as many as 1,200 a year over their ten-year lifespan. As if that weren't enough, giant African land snails can store their mate's sperm for up to two years, fertilizing multiple batches of eggs. It all adds up to some serious potential for invasiveness.

The problem is particularly bad in Florida because the snails have no natural enemies there. Plus, their colossal size, which allows them to muscle out native species of snails, isn't helping either. They can eat pretty much any plant material, over five hundred species in Florida alone, making them a nightmare for gardeners or really anyone who appreciates having the slightest hint of green in their yard. And that giant shell, which can puncture car tires and breaks into biological shrapnel when hit with a lawn mower, isn't going to build itself, so the snail devours any source of calcium it can find. Thus, in addition to ruining your tomato plants, the snail will turn around and lay siege to your house, gnawing off the calcium-rich stucco, which is made partially of lime, aka calcium oxide. And it even attacks *concrete*. If kept in captivity and not provided with enough calcium, it'll eat the shell off another snail's back. (For this the snail's equipped with a tongue of sorts, known as a radula, which is packed with tiny, tough teeth. So the snail doesn't so much chew up plants and houses as it does rasp them away bit by bit.)

The situation in Florida is dire, but it ain't the state's first rodeo with the invasive snail. Back in 1966, a family went on vacation in Hawaii and returned to Florida with a few extra members: The son smuggled back three giant snails and handed them over to his grandma, who released them in her backyard. The population exploded and spread. A decade later, officials had destroyed some eighteen thousand snails at a cost of a million bucks. The giant snails were vanquished—for the time being.

Fast-forward to September 2011: the Second Coming of the Snails. It was a Miami homeowner who first discovered them, and a mere six months later officials had collected forty thousand snails, more than twice what they'd found over the decade of the first invasion. Three years later, that figure had ballooned to 140,000. In fairness, the government has done a fine job reining the things in, keeping the giant snails mostly contained in the Miami-Dade area. They've mounted a huge public outreach campaign, complete with old-timey WANTED posters that bear photos of the pest and a hotline to report sightings, though it should be noted that they lack any mention of a REWARD. But reward notwithstanding, the campaign seems to be working, and it's a good thing it is. At stake is billions of dollars' worth of agriculture statewide. If the giant snails start radiating out of Miami-Dade, it'll be a disaster. Pretty much everything is on their menu, including that famous Florida citrus we're all so fond of.

FIRST STOP: SNAILS. NEXT STOP: CHEAP PARTICLEBOARD FURNITURE

The giant African land snail may not have predators in Florida, but in Brazil, a bird called the great antshrike has struck on a novel way to control the invasive species: It grabs the juvenile snails and carries them back to its favorite rock and repeatedly smashes them against it. It's a fascinating case of tool use among birds (crows have been known to use much bigger tools for cracking open food, placing walnuts in the middle of the road for cars to run over), and it's pretty impressive, if you ask me. I mean, I've got a great big human brain and I still struggle to build IKEA furniture, even with those nice little illustrations in the instructions.

I can tell you who's not helping: folks who are fond of certain rituals, and who may have been responsible for Florida's Second Coming of the Snails in the first place. A man allegedly smuggled the snails in and raised them in a box in his backyard, before slicing them open while they were still alive and pouring their bodily juices into his followers' mouths. The expected ensued. His loyal subjects fell violently ill, "losing weight and developing strange lumps in their bellies," according to the *Miami Herald*. But they could have had it much worse. The giant African land snail carries the dreaded rat lungworm, which infiltrates your brain, paralyzing or blinding or even killing you.

Thus are the exploits of the giant African land snail, coveted by a few goofballs and dreaded by just about everyone else.

Aye-Aye

PROBLEM: Grubs are packed with protein. But the thing is, grubs can pack themselves into the safety of a tree branch.

SOLUTION: A haunting lemur called the aye-aye develops beaverlike teeth to gnaw into the wood, plus a freakishly long finger to fish out grubs.

Eight months before Darwin published *On the Origin of Species*, the famous English anatomist and accomplished curmudgeon Richard Owen got a letter from a countryman posted as colonial secretary on the island of Mauritius. Before he'd left England, the secretary promised to send Owen any interesting specimens of natural history, and he was now writing to say he had a doozy from the neighboring island of Madagascar: an aye-aye, named after the natives' cries of astonishment when glimpsing one. It was a bizarre kind of lemur, about the size of a cat, with huge ears, "teeth as large as those of a young Beaver," and a middle finger "being slender and long, half the thickness of the other fingers, and resembling a piece of bent wire." The secretary added with a nudge-nudge-I-really-went-out-of-my-way-for-you that "the Aye-aye is an object of veneration at Madagascar, and that if any native touches one, he is sure to die within the year; hence the difficulty of obtaining a specimen. I overcame this scruple by a reward of £10." *Nudge nudge.*

The secretary put the aye-aye in a wooden cage and dropped in some tree branches bored through with grubs. These interested the aye-aye. The creature examined the branches, bent its big ears toward them, and "rapidly tapped the surface with the curious second digit, as a Woodpecker taps a tree, though with much less noise, from time to time inserting the end of the slender finger into the worm-holes as a surgeon would a probe." It was tapping to agitate the grubs, and listening for the resulting movement. The worms would retreat deeper into their burrows, but eventually the aye-aye succeeded in catching them, gnawing through the bark into a chamber and reaching in with that long, slender finger to haul out the prize.

On such a diet (supplemented with dates—for fiber, I guess?) the secretary was able to keep the aye-aye alive. But the animal eventually escaped, whether by gnawing through the wooden bars with its beaverlike teeth or some other means isn't clear from the secretary's letter. It was soon recaptured, though, and, per Owen's instructions, "killed by chloroform, its arterial system injected, the cranial cavity exposed, the abdominal cavity and ali-

MIGHT AS WELL JUMP

Over on the islands of Southeast Asia, a different primate has modified its fingers for other means. The tarsier is a tiny nocturnal creature that would fit in the palm of your hand, but it's a voracious hunter, bounding from branch to branch snagging insects. Its fingers are slim and elongated, much like the aye-aye's, but topping each one is a fleshy pad that helps the tarsier stick landings. Which is just as well, because in addition to its modified hands, it has a highly elongated tarsus—the bones that make up the ankle and give the tarsier its name—that allows it to bound an incredible fifteen feet. That's the equivalent of a six-foot-tall human leaping 180 feet. Doubt you'd be able to stick the landing, though.

mentary canal injected with alcohol, and the whole animal then immersed in a keg of colourless spirit." The secretary then shipped it to Owen, who proceeded to do a seventy-page write-up on the anatomy of the Malagasy curiosity.

Now, Owen has been cast as Darwin's nemesis, a reactionary defender of Victorian ideals and a combative old man who loathed the idea of natural selection and above all else believed that a higher power had tailor-made every creature for its niche. And it probably hasn't helped his reputation over the years that he really was a bit of an ass, or that the guy looked real sketchy, like the love child of the Wicked Witch of the West and Ebenezer Scrooge. (Conveniently enough, one of Darwin's allies who publicly attacked Owen, Thomas Henry Huxley, branded him a "humbug." Huxley got all up in Owen's grill, so to speak, in a legendary meeting of scientists in which Owen claimed humans were distinct from gorillas, and therefore could not have descended from apes, because only human brains had a hippocampus—except that's not strictly speaking true at all.) And it didn't help that when Darwin published *On the Origin of Species*, Owen anonymously penned a scathing review, a stunt that Darwin chalked up to jealousy in his autobiography.

In fairness, though, Owen had his own vague, undeveloped evolutionary ideas, however much he savaged Darwin and his allies. And ironically enough, one of Owen's greatest theories— which predated Darwin's discovery of natural selection—lent huge insight into evolutionary theory. Owen realized that something like the aye-aye's hand wasn't just a modified primate hand, but a modified *mammal* hand, because a creator had used the same blueprint, which the cranky anatomist called an "archetype," across all manner of creatures. He had discovered what Darwin later realized was a component of common descent, that an aye-aye hand is like a human hand because long ago we shared an ancestor that had such a thing. The aye-aye went one way and developed that elongated finger, while we settled on our more uniform digits.

THAT'S A BIT OF A STRETCH

Another debunked evolutionary theory of note is that of French naturalist Jean-Baptiste Lamarck, who prior to Darwin had argued that animals could acquire traits during their lifetimes and pass said traits to their children. So giraffes got their long necks by stretching to reach the tops of trees, stimulating the concentration of a so-called nervous fluid to encourage growth. In reality, mutations that produced longer-necked giraffes allowed the creatures to exploit out-of-reach leaves, boosting their chance of survival and therefore their chance of passing along long-neck genes. But to his credit, Lamarck also posited that organs that creatures no longer have much use for fade away over time, which is entirely true, as in the case of the naked mole rat's atrophied eyes.

But why? Well, the same forces that made the satanic leaf-tailed gecko so satanic and so leaf tailed. When Madagascar set out on its own, its ecosystem went into flux. In their isolation, the island's animals have assumed any available niches—the aye-aye, for instance, taking on a role hunting wood-boring grubs, a role that would typically go to a woodpecker. (On the other side of the planet, in Barbados, another curiosity, called the threadsnake, has taken on an ant-hunting role that'd typically go to an invertebrate, a centipede maybe, shrinking so much over evolutionary time that it can now curl up on a quarter. That's how powerful of a force isolation can be.) With its conspicuous lack of a beak to chisel away at tree branches, the aye-aye has had to make do with the modified tools of a primate: those dexterous fingers, and fused, rodentlike teeth so tough that aye-ayes in captivity have been able to chew through cinder blocks. How dexterous are those digits? The aye-aye's middle finger actually has a ball-and-socket joint like a shoulder, allowing the digit to swivel around as it feels through the grubs' chambers.

The aye-aye is as good a demonstration as any of the awesome power of natural selection and the interrelatedness of all living things. Going even further back than the ancestor we share with the aye-aye, hundreds of millions of years ago fish dragged themselves onto shore with almost handlike fins, structures that may have evolved to help them escape from predators in the water. Those fins were a blueprint. The fish evolved into a wide variety of tetrapods (meaning "four-footed")—all land-bound vertebrates like the amphibians and reptiles and mammals and birds—but the blueprint remained. The sociable weaver's fin is a wing, the antechinus's a paw, ours a hand, and the aye-aye's a hand superspecialized for capturing grubs, because we're all descended from a common fish ancestor.

Owen was a piece of work but he was right, however accidentally right he may have been: There's a blueprint, and a beautiful one at that, which the automatic forces of natural selection have modified time and time again as a solution to certain problems. You and I, Ebenezer Scrooge and the aye-aye, the Surinam toad and the pink fairy armadillo, we're all that same fishy design customized for different purposes, for it's the limb that binds us.

Mantis Shrimp

PROBLEM: All kinds of seafloor critters have robust armor, thus requiring the right utensils to eat them.

SOLUTION: The mantis shrimp deploys hammer-hands that strike its prey so fast they momentarily heat the surrounding water to the temperature of the sun's surface, splitting clams and crabs to pieces.

Biologist Lindsey Dougherty specializes in the disco clam because, as far as clams go, it's pretty amazing, with its gaudy, rapidly flashing bands that run along its mantle. At first glance you might think that the creature is bioluminescent like the anglerfish's lure. But instead of producing light, this is special tissue that's so reflective that it fires bright light at your eyeballs. Dougherty had a hunch that the show was meant to scare away predators, so she got a creature so fearsome, so ornery, so powerful, that you'd expect the clam to flicker like it never flickered before.

She got herself a shrimp.

Okay, maybe that's a bit misleading. Even though it kind of looks like one, the mantis shrimp isn't a shrimp, but another crustacean known as a stomatopod. But all that other stuff is true—these things are mean, immensely powerful, and not to be messed with. There's a reason they're called thumb splitters.

As Dougherty's camera rolled, the stomatopod approached the disco clam, grabbing it and sizing it up. But the master predator suddenly recoiled. And approached again. And recoiled again and approached again and recoiled and then for some reason went catatonic, before snapping out of it and proceeding to hump the clam. I'd like to think that the stomatopod was so frustrated by the fact that it had come across something it couldn't murder that it could no longer cope with reality, but in actuality it probably had more to do with the sulfur in the disco clam's flowing tentacles having some sort of effect. Still, stomatopods are not used to giving up. If something is a reasonable size to attack, chances are a stomatopod is equipped to destroy it. This business with the disco clam, it was a fluke.

There are two types of stomatopods, the spearers and the clubbers, distinguishable by the weaponry that sits below their face. The mantis shrimp get its name from the first group: It has two spiked arms that look much like those of the famous insect, only flipped upside down (it snags fish from below, grabbing them like

---

IT'S ONLY OKAY TO ATTACK KITTENS THAT ARE EVIL

Crustaceans like the stomatopods get typecast as marine creatures, but there are plenty of terrestrial varieties. Your average pill bug—or roly-poly, or whatever you like to call it—is in fact a crustacean. And then there's the mightiest land crustacean of all: the coconut crab. It's a kind of hermit crab, only it forgoes the shell and grows to three feet across and ten pounds. The crab's claws are so powerful it can tear through coconuts, as its name suggests. On the islands in the Pacific the coconut crab calls home, it's been known to attack . . . kittens. I'm sorry you had to read that, but who knows, maybe those cats were evil or something. It's good to think positive, after all.

you'd carry a bundle of sticks on your forearms). This kind tends to dig burrows in the sand, lashing out at passing fish so quickly that the strike is almost invisible to our eyes, before disappearing back into its lair, leaving only a shimmering cloud of scales for evidence. It's so fast, in fact, that the world expert in the mantis shrimp's strike, Sheila Patek, couldn't at first film it because her university's cameras weren't able to capture enough frames per second. Only when a BBC crew showed up did Patek get her footage: She convinced them to rent a brand-new camera model powerful enough to pull it off—in exchange for giving them a story, of course.

But the strikes start bordering on the far-fetched with the clubbers. At work here is a surprisingly simple mechanism. Up at the top of the arm is a kind of membranous divot, curved like a saddle, that acts as a spring. The stomatopod contracts a big muscle in that arm, pulling back the club and compressing the spring until an internal latch snaps into place. That latch is holding back a tremendous amount of energy in the muscle, and when it flies open, the spring rockets the club forward at up to fifty miles per hour, which is especially impressive considering this is happening in water. The resulting force is devastating. Subjected to such treatment, clams shatter, while more fragile creatures like crabs explode into a shower of limbs. Indeed, a clubber stomatopod will target a crab's claws first, blowing them off to counter the prey's futile attempts to defend itself.

Here, too, Patek realized that the stomatopod confounds human technology. When she went to measure the forces the clubber could produce, she found that the instrument she'd bought for the creature to smash was inadequate. It was rated for forces up to one hundred pounds, when it turns out the stomatopod can punch at well over two hundred pounds. When she finally got her readings with a different instrument, Patek noticed something strange. There was the initial force, but also a second impact that hit a fraction of a second later: cavitation bubbles. Like the pistol shrimp, the clubbing stomatopod is striking so quickly that it creates these supremely destructive bubbles, which spread over the surface of the prey and explode not just with a shock wave and temperatures almost equal to those on the surface of the sun—nearly 10,000 degrees Fahrenheit—but a flash of light. The strike is a one-two punch, an initial impact with a trailing explosion.

You'd expect all the smashing to take a toll on the weapons themselves, and it certainly does. Patek has seen stomatopods wear their clubs down to the muscle (again, these things are plain ornery, not to mention hungry). But by virtue of being an arthropod—that is, an invertebrate that among a few other defining characteristics has an exoskeleton—it can shed that armor and grow it again, complete with shiny new clubs. Sure, there's a bit of a perilous period there as it waits for its shell to harden up, but that's nothing a little bluffing can't fix: When threatened during

THE ANTS THAT WOULD BE POPCORN

Sheila Patek has become the go-to scientist if you're looking for information about the fastest, most powerful strikes in the animal kingdom. Besides the mantis shrimp, her other interests include the trap-jaw ants. Like the clubber stomatopods, they use a latch mechanism to cock their mandibles and store up tremendous energy. Special hairs on those mandibles trigger the strike when they brush up against prey, crushing or launching the stunned victim through the air. And when threatened, trap-jaw ants even turn their weapons on themselves. By pointing their mandibles at the ground and firing, the ants fling themselves away from danger, tumbling end over end. Disturb one of their nests and they'll do it en masse, popping off like angry popcorn.

this time the stomatopod won't smash, but instead waves its arms around and hopes that's enough to deter predators.

Really, it'd be hard to overstate what an amazing tool the exoskeleton is for the stomatopod and the other arthropods that swim or fly or walk the Earth. It wins those mantis shrimp meals and keeps them from blowing their own arms off, but it also provides protection from predators and the elements. And as we'll see in a bit, the exoskeleton has played no small part in helping the beetles, of all things, conquer the Earth. The shell can take on all manner of colors for advertising sexiness to mates or poisonousness to hungry enemies. It can form into bizarre structures, while we mammals are pretty much stuck with these fleshy vessels.

That's not to say we're chopped liver. Those exoskeletons keep arthropods from growing too big, since they start getting prohibitively heavy at a certain point, whereas vertebrates can grow enormous with their endoskeletons as support. It's just that evolution has gifted invertebrates and vertebrates with different strategies. We got our great big brains and the stomatopods got their weaponry. And that, quite frankly, is fine by me. Being strong enough to punch through a clam is a lot of responsibility, after all.

Bone-Eating Worm

PROBLEM: The deepest of seafloors are desolate, and food is in short supply.

SOLUTION: A certain worm has hit upon the idea of boring into the skeletons of creatures that have sunk to the bottom, dissolving and digesting bone with the help of friendly bacteria.

Creatures are dependent on sunlight in surprisingly complex ways. The organisms with the first dibs on sunshine are the plants, which use it in photosynthesis. The sun's energy is then passed up the food chain, to the bugs that eat these plants, to the birds that eat the bugs, to the mammals that eat the birds.

The same goes in the oceans. Plantlike phytoplankton float around absorbing the sun's energy. Zooplankton eat the phytoplankton, and fish eat the zooplankton, on up the line. Every critter in that food chain, though, meets its end and sinks down into the abyss, a biomass that's known quite tranquilly as marine snow—even though it's made of dead things. Opportunists in the water column pick at this marine snow as it falls, and accordingly very few nutrients even reach the seafloor. So few nutrients, in fact, that a whale that dies and sinks more or less intact to the bottom (there may be opportunists, but not enough to strip a whale clean before it hits the floor) will provide as much food to the critters of the seafloor as would thousands of years of marine snow.

The so-called whale fall is what the scavengers of the seafloor are desperate for. They'll pick the giant's bones clean and won't stop at that, for there are creatures that devour the bones themselves.

Worms of the genus *Osedax* are quite beautiful, being one or two inches long with a white tube of a body, at the end of which erupt red or pink or orange frills, known as palps, which gather oxygen. But it's what you can't see that's so amazing about the bone-eating worms. Down inside the whale's bones are worms' guts. Well, gutlike organs. The worms don't have mouths and they don't have intestines. Instead, they send "roots" down into the skeleton, forming a structure that varies between a simple bulb and thinner, wandering shoots, depending on the worm. These

OSEDAX: MUCKING UP THE FOSSIL RECORD FOR 100 MILLION YEARS

Getting fossilized is a pretty tall order. The conditions have to be just right—dying in sediment helps—but even then you have to hope that scavengers don't cart you away in pieces first. Soft-bodied creatures like worms are especially resistant to fossilization, but that doesn't mean they don't leave behind evidence of their existence. For instance, scientists have found the fossilized bones of sea turtles and plesiosaurs (those fearsome, long-necked marine reptiles) bored through with holes, showing that squishy old *Osedax*, having resisted fossilization itself, has been mucking up the fossil record for at least 100 million years. Excuse me. The bone-eating worms, in the words of the scientists, "may have had a significant negative effect on the preservation of marine vertebrates in the fossil record." Yes, that sounds much more scientific.

roots release massive amounts of acid, while inside are symbiotic bacteria, which take the fats and proteins absorbed through the worm's tissue and convert them into energy. All the worms have to do is sit there—oftentimes in vast sheets covering the skeleton, swaying gently with the current like tiny trees sucking up groundwater—as the bacteria go to work. In exchange, the bacteria get themselves nice little homes.

Surprisingly enough, though, eating bones isn't a big deal. Hyenas do it with their powerful jaws, biting off bits of ribs like you or I would eat sticks of celery. It's *how* the bone-eating worm does it that makes it so unique in the animal kingdom—no other animal feeds in this way—not to mention how dependent it is on the strategy. Bones for a hyena are dessert after a main course of flesh, but for the bone-eating worm, skeletons are all it has. On the menu is liquefied bone, always and forever. And the worm is lucky it even has that. It isn't every day that a whale dies and sinks to the bottom of the sea, and it's certainly unlikely that two will fall right next to each other. So the bone-eating worm stakes its claim, permanently anchoring itself into a whale bone and chowing down.

Such immobility would seem to present a problem for mating. Even more problematic, scientists at first were finding only females on the whale bones. So where were the males? It wasn't until researchers opened the worms up that they found microscopic males *inside* the females, up to one hundred of them, about one hundred thousand times smaller than their mates. Really, the males are simply sacs of sperm and yolk. Unable to feed themselves, they live off the yolk afforded to them at birth, producing sperm as their food supply dwindles, until their seed is gone, their sustenance is gone, and their body cavity is empty.

But how do the males find the females? Well, scientists reckon that bone-eating worm larvae float around on currents and will develop into males only if they land on a female. Those that land on bones anchor themselves and become females, developing into that lovely worm and waiting for a lucky soon-to-be male to arrive. When that happens, their gametes will mix and the female

DYING TO GET LAID

Sexual internment sounds like a raw deal, but it could be much worse for the male *Osedax*. Some fellas in the animal kingdom pay for sex with their lives—and by that I mean the females eat them. One theory for why this happens posits that by eating the male, the female gains valuable nutrition to put toward developing her young. Thus, the male has a better chance of passing his genes along even though he's dead. Indeed, one study of praying mantises showed this, with females in crummy shape being more likely to snack on their mates than their healthier counterparts. The cannibals in turn increased the weight of their egg cases . . . and significantly reduced the odds of the male ever calling back.

will release her young into the current, thus spreading the worms around the seafloor. It all may sound overly complicated, but it makes good evolutionary sense. Males that don't feed don't need to worry about competing with females for the scant food on the seafloor. And by permanently attaching himself to a female, the male all but guarantees he can pass his genes along. The whole thing is remarkably similar to the goings-on of the sex-changing tongue-eating isopod, not to mention the anglerfish: Males never eat, and when they find a lady, they'll be damned if they're going to let her slip away.

Strangely, though, there's one particularly tiny species of bone-eating worm, *Osedax priapus*, that's evolved away from this mating system. Its males are much closer to the size of the females—they're only about three times smaller, as opposed to one hundred thousand times smaller. Males don't attach to females, but instead stretch their bodies to hand sperm off to their neighbors. So why deviate? The reason could come down to *Osedax priapus*'s small size. Males in other species of bone-eating worms may have evolved into dwarfism because of competition over food: These things totally swarm whale skeletons, so if the male doesn't need to feed, all the better. *Osedax priapus* is so tiny, competition to feed on a given bone might never have been a problem. The clingy dwarf *Osedax* males may have a guaranteed mate for life, but the large males of this particular species can gorge themselves on bone, gaining more energy to produce sperm, plus they can mate with multiple females instead of committing to one.

So in the end, it comes down to food provided by the sun. Zooplankton eat the solar-powered phytoplankton, and the whale at the top of the food chain eats the zooplankton, before perishing and falling and serving itself up to the creatures at the very bottom. An *Osedax* worm finds a bone, and an *Osedax* larva finds an *Osedax* lady, and the rest is a story for their grandkids.

Wait, that's not a good story to tell the grandkids. Disregard that.

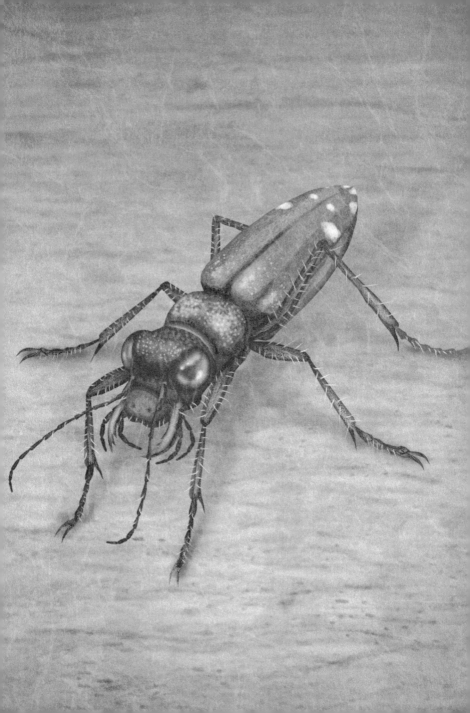

Tiger Beetle

PROBLEM: Your prey are quick footed.

SOLUTION: The tiger beetle one-ups them by running so fast it blinds itself and has to stop every once in a while to get its bearings. Not that it matters. This is a supreme sprinter with a stomach to fill.

In October 1858, four years into his travels hopping between the many islands of Southeast Asia, Alfred Russel Wallace found himself in a clearing in a forest, where he apparently interrupted some kind of grand *Snow White*–esque meeting of animals. Specifically, beetles: There were weevils and longicorns and gorgeous golden varieties—so many beetles, he recalled, "that they rose up in swarms as I walked along, filling the air with a loud buzzing hum." As any self-respecting naturalist would do, he returned to the clearing for the next three days, collecting some one hundred different species. All in a single jungle clearing.

Here's a figure for you: one-fourth. That's not the fraction of Earth's species that are invertebrates. That's not the fraction of Earth's invertebrates that are insects. That's the fraction of all animal species on Earth that are beetles. One. Quarter. So perhaps it's not surprising that Wallace stumbled upon such a bounty of the bugs, for beetles are unparalleled in their success in the animal kingdom. Some are built like tanks. Some specialize in rolling dung into balls and devouring them because, hell, someone has to. Still others are nimble, turbocharged killers, and king among these are the comically long-legged tiger beetles, insects that chase their prey with such speed that they go blind.

Now, it wouldn't be quite fair for me to just tell you that among the 2,700 species of tiger beetle, there's a particularly speedy Australian variety that's been clocked running at 5.6 miles per hour. That seems like nothing. Until you consider that according to the British Heart Foundation, when walking for exercise, four miles per hour is recommended "for a person with excellent fitness," while three miles per hour is more average. So the next time you get up and take a few steps, know that a beetle three-quarters of an inch long could have burned you. If it were running alongside you for one second, the tiger beetle would have covered 125 body lengths while you covered less than one of your own. Even at full gait, the fastest humans can cover only six body lengths a second,

---

WINGING IT

So what gives with all the success for the beetles? A lot of it comes down to their characteristic wing covers, known as elytra, which are modified wings that shield their flight wings. You know the spotted coverings on ladybugs that snap open before they take flight? Those are the elytra. For species of beetles that have lost their ability to fly, the elytra function as extra armor. For beetles with a penchant for water, the coverings trap air bubbles, so the creature can breathe while submerged, a lot like the diving bell spider. And desert-dwelling species rely on the elytra to help retain moisture. So think of the elytra as a superhero's cape of sorts, only instead of being made of fabric they're made of "highly sclerotized dorsal and less sclerotized ventral cuticles." Which sounds much more impressive, to be honest.

while cheetahs manage sixteen. For its size, the tiger beetle is an incomprehensibly fast animal, hitting speeds equivalent to a human being sprinting at 480 miles per hour.

The beetle is so fast, in fact, that even though it has some of the sharpest eyes among insects, its giant peepers can't collect enough light when the creature is in hot pursuit of prey. So every once in a while the tiger beetle has to stop to again lock onto its quarry. In an average pursuit, this will happen three or four times, but the beetle is so fast, the setbacks don't matter a bit. It catches up with the prey, snagging it with enormous mandibles and ripping it to pieces.

But it's not only the prey that the tiger beetle needs to keep its eyes on during these pursuits. There's also the matter of avoiding rocks and sticks and such, for a good face-plant can really set a predator back. It turns out that it isn't all about vision here: The beetles are feeling their way around by holding their antennae forward, with the nocturnal species among them tending to kind of twirl them around in circles (cockroaches, which are also largely nocturnal, do the same to sense their world) and diurnal (a five-dollar word meaning daytime) species holding them straight out and pointed slightly downward. That's interesting because while we'd expect nocturnal insects to rely on their antennae and not their eyes to make sense of their world, the diurnal varieties of tiger beetle are so fast they have to supplement their vision with mechanosensation—that is, by picking up mechanical cues from the environment.

You might be wondering how scientists figured this out in such a fleet-footed creature. It was pretty simple, really. Researchers got three groups of tiger beetles: one normal, one whose eyes they painted over, and one whose antennae they cut off. They then put together an obstacle course and set up a camera shooting four hundred frames per second, and with a few encouraging bumps with a paintbrush, sent the beetles scurrying. Beetles with their antennae amputated did a whole lot of face-planting, but the normal and blinded beetles performed equally well. The video showed that the antennae are pivotal for obstacle avoidance: Once

A GENERALLY ELONGATED CREATURE

Arachnids don't have antennae, but a particularly creepy group among them sure think they do: the whip spiders. (Not technically spiders, and don't confuse them with the similar-looking whip scorpions, which aren't technically scorpions, though whip spiders are also sometimes called tailless whip scorpions. Make sense?) Like the tiger beetles, they feel their way around their environment, only they do so with their absurdly elongated front legs, which can be four times as long as their other already absurdly elongated legs. When one of those front legs hits a victim, the whip spider snags it with spiny graspers called pedipalps and gnaws on it alive, adding digestive juices and slurping down the resulting smoothie. (Not technically a smoothie.)

the feelers made contact with the barrier, they would briefly snag and bend before snapping back into place as the beetle detected the impediment and bounded over. So out in the wild, even if it's a diurnal hunter, the tiger beetle is behaving more like a nocturnal one, its vision snatched away not by darkness, but by raw speed.

Even tiger beetle larvae, which are largely sedentary and don't yet have the lanky legs of their adult form, are in their own way as blindingly fast and brutal as their elders. They hang out in cylindrical burrows in the ground, with their armored head and giant mandibles plugging up the hole, while their elongated body—which has two hooks on the back to anchor the larvae so they don't get dislodged when they've got hold of larger prey and it's trying to wiggle free—extends below. If an unfortunate insect like an ant happens to stumble by, the tiger beetle larva will fire part of its body out of the burrow, snag it, and yank it down into what you can imagine is a terrible death. Should it overwhelm enough prey, the larva will grow big and strong, leaving behind its cozy burrow for the life of the world's fiercest, most remarkable sprinter.

CHAPTER 7

You Can't Let Them Get Away That Easily, Can You?

In Which Predatory Snails Weaponize Insulin and Bugs Have the Nerve to Assault Charles Darwin

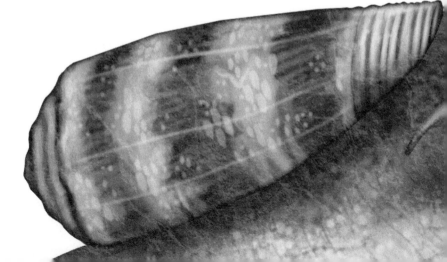

Being able to run down prey is one thing, but keeping a grasp on them is a whole different challenge. For that, evolution has come up with some pretty inventive solutions. I'm talking poison arrows. Slime cannons. And a particularly nasty mouth that had the gall to attack Charles Darwin himself.

Bolas Spider

PROBLEM: Moths can escape typical spiderwebs.

SOLUTION: The bolas spider mimics a female moth's sex pheromones, luring in the males. Then it swings a specialized web, just a drop of goo on a line, to snare its prey.

In his South American travels, Darwin developed quite an affection for the gauchos—the masterful horsemen and southern cowboys—who stunned him with their skills. Among their tools, none was more effective than the bolas, two or three stones or iron balls wrapped in ropes of leather to create twirling mayhem. When gauchos threw the bolas at prey, the whole mess would tangle around limbs, oftentimes with enough force that it'd snap bones.

Ever the good sport, Darwin figured he'd give it a whirl. Galloping about on his horse and spinning the bolas above his head, he let loose on . . . himself. The bolas hit a bush and dropped to the ground, entangling in the steed's limbs. "Luckily he was an old practiced animal," Darwin writes in *The Voyage of the Beagle*, "and knew what it meant; otherwise he would probably have kicked till he had thrown himself down. The Gauchos roared with laughter; they cried out that they had seen every sort of animal caught, but had never before seen a man caught by himself."

Appropriately enough, sitting in the very bush that Darwin assaulted may well have been an arachnid that takes its name from the gauchos' mighty weapon: the bolas spider. This kind of orb weaver isn't spinning complex webs or throwing leather-wrapped balls of iron at its prey, but it is hunting them in a similar, though far more complex way. During the day, bolas spiders hunker down in plain sight, with some species conveniently resembling bird crap, thanks to a splotchy black-and-white pattern that makes them look unpalatable. But at night, when the spiders emerge to hunt, the males and females take up very different strategies.

The female bolas forgoes all the work it takes to construct an intricate web—instead, she extends a single line of silk between two leaves or twigs. To that she attaches one more line. As she's spinning the second line, she uses her hind legs to comb a sort of glue from her spinnerets onto the silk, forming a viscous glob at the end. When she cuts the line off from her spinneret, the weight of the glob pulls the line down, so it hangs vertically, perpendicular to the supporting silk. Her weapon at the ready, she holds it with a leg and waits.

Depending on the species, the spider uses her bolas in different ways to hunt moths, her prey of choice. On one end of the spectrum, she'll wait until she's detected a target before she flings the bolas with great speed and precision, snagging the moth with the glue. (It's so fast, in fact, that one species, *Mastophora dizzydeani*, is

THE MERITS OF LOOKING LIKE BIRD TURDS

Lots of animals disguise themselves as bird crap because, let's face it, no one wants to mess with bird crap. Perhaps the most spectacular among them are some of the bolas spider's relatives in the orb weaver family. Unlike the bolas, these do indeed spin intricate webs, but that leaves them out in the open and vulnerable to predatory wasps. So orb weavers spin thicker white silk around the center of their web that looks like a splat, then hang tight there, with their splotchy body coloration completing the ruse.

named after the baseball pitcher Jerome "Dizzy" Dean, who was so called because he was an eccentric, not because he had balance problems. As for the name "bolas spider" in general, one researcher advised that the name not be changed even though it doesn't work *exactly* like a bolas, but he did suggest that, no joke, a more accurate moniker would be "sticky yo-yo spider.") Still others take a more spastic approach, rapidly whirling the bolas when a moth is nearby until the thing runs into the trap. And still others don't even bother waiting for the prey to show up, flailing about as soon as the line is ready for as long as fifteen minutes. That glue will start evaporating away, so if she's unsuccessful in her hunting after about a half hour, the spider will gobble up the line.

She's going through all the trouble because moths can escape your typical spiderweb. They'll stick, all right, but they're covered in scales that rip off and allow the moth to escape. Accordingly, the bolas spider's trap is no simple dab of glue. The goop is made up of two liquids of differing viscosity. The outer layer is less viscous, while the inner layer is nice and thick. This inner layer also holds a mass of extra coiled silk, allowing the glob to elongate as the spider swings the line around, increasing the strike distance. When she does snag her prey—and she most certainly will, an average of twice a night—the less viscous outer layer flows past the moth's scales and sticks to the cuticle below. And that extra silk in the glob acts as a shock absorber, both for the initial impact and as the moth struggles to free itself. Prey secured, the bolas spider climbs down the line, gives the moth a bite, and starts wrapping it up, sometimes leaving it right there on the thread as she builds out new bolas.

Strangely, though, the bolas spider hunts only the males of specific moth species. How could this be? How could the bolas snag only particular kinds of moths when the forest is swarming with insects, and for that matter, why would she want to limit herself to a tiny menu? Well, it turns out the bolas spider is releasing scents that mimic the sex pheromones of female moths of a given species. Male moths swooping in thinking they're going to get laid instead

THE MERITS OF HAVING A GLOWING BUM

Even though it's unrelated to the bolas spider, there's a glow-worm in the caves of New Zealand that's hit on a similar manner of hunting its prey. It's the larval form of the fungus gnat, and like the spider it begins by stringing a horizontal support line, in this case a tube of silk filled with mucus. From that it hangs up to seventy fishing lines, each loaded with several droplets of glue. But instead of giving off pheromones to attract its prey, the larva glows with bioluminescence. When an unfortunate insect homes in on the light and slams into the line, the hunter squirms through the mucous tube, reels up the catch, and devours it. So, unlike the bolas spider, it spares itself the weirdness of having another species sexually attracted to it, which is nice.

find themselves tangled in the bolas spider's trap. The spider can even mimic the scents of two different kinds of moths with two different pheromones by putting off a smell that's a mixture of the two. The bolas spider *Mastophora hutchinsoni*, for instance, targets a moth that appears early in the evening and one that shows up later. By modifying her pheromone as the night wears on, lowering its attractiveness to the first moth species, she can better attract the latter.

Such specialization poses a problem. While other spiders build their webs and catch insects indiscriminately, the bolas spider can find herself in a habitat without any of her target prey. So she has to test the environment. When night falls, before she even begins building her bolas, the female emits the scent that drives male moths nuts. If none show up, fine: Instead of wasting the energy and resources and time building a bolas, she moves on to another part of the forest. But if the moths do show, only when they get close enough for her to feel the vibrations of their wing beats does she start setting up shop. (If you have a mind to prank a bolas spider, just hum near her and she'll think you're a moth and start building a trap. Seriously.)

As for the males, they're comparatively tiny—remember the size rule with spiders. Being so small, they can't take on moths, so they don't even bother building bolas. Instead, they wait at the edges of leaves and use their hairy arms to snag flies. That's a whole different niche, keeping the male and female bolas spiders from competing with one another for food. The male doesn't get to bolas him some moths, sure, but then again, unlike Darwin he'll never know the emotional pain of screwing it up and having all his friends laugh at him.

Velvet Worm

PROBLEM: Worms aren't celebrated for their speed.

SOLUTION: But the velvet worm *is* celebrated for its weaponry. It fires jets of glue out of two modified legs—yes, legs—to entrap its prey. Then it takes its time gnawing through their exoskeletons.

The bolas spider enlists a complex system of chemical seduction and sticky entrapment, but you might say it's living in the dark ages. After all, having to be so close to your victim to launch an attack is like a knight risking injury to draw in close and stab his victim (chemical seduction optional). So if the bolas spider is the foot soldier of yore, surely the velvet worm is a deadeye gunslinger. Like the bolas, it deploys its own immobilizing goo, only from a distance. The worm then leisurely closes in on its prey—no hand-to-hand combat required.

The first thing you need to know about the one hundred or so species of velvet worm, which grow up to six inches long and keep to tropical and temperate forests, is that they're pretty much all legs, even the bits that don't seem to be legs. Depending on the variety, there can be as few as a dozen pairs of the stumpy, squishy little things or as many as forty, all supported with pressurized fluid. But the legs don't stop there. In a clever evolutionary maneuver, the velvet worm has adapted a pair of legs into antennae, another into slime cannons, and yet another into jaws. That last one shouldn't make a lick of sense, but it turns out that each leg is tipped with a claw that helps the creature get a grip in the undergrowth—indeed, the worms are known collectively as the ony-chophorans, or "claw bearers"—a great material to turn into a pair of jaws, each of which is a pair of fanglike blades.

Those leg-antennae go a long way in helping the velvet worm sniff out its prey's chemical cues, but the hunter's most improbable sensory organ is its entire body. Covering the worm are minute bumps that calculate shifts in air currents from the movement of potential prey, including spiders and termites, on up to larger

LEGGING IT

The many-legged velvet worm may resemble the centipede or millipede, but it's only distantly related to these armored arthropods. The venomous centipedes, the name meaning "hundred feet," usually have far fewer than one hundred legs, and the harmless millipedes, with their supposed one thousand legs, top out at 750.

If you come across something you suspect is a centipede or millipede and feel like picking it up, a good way to tell the two creatures apart is if it assaults you. Or, better yet, don't let it come to that. The more cylindrical millipede is a sluggish and harmless detritivore, meaning it feeds on decaying plant matter. The centipede, however, is a streamlined and frenetic hunter. Like the velvet worm, it has modified two of its legs into weapons, in its case claws that deliver venom. Those toxins may work magic on smaller foes, but for us humans the effects are typically no worse than from a bee sting (though—as with bee stings—some folks may experience a more severe allergic reaction), and that's even if the claws can manage to penetrate the skin. But don't say I didn't warn you.

quarry like beetles and crickets. When the worm does lock on to something, it takes its sweet time to sneak up on the soon-to-be victim—partly because, hey, it's a worm, and it can't be rushed—and partly because you don't need speed when you've got cannons on your face.

To the naked eye, the blast looks simple. Poke a velvet worm and there's suddenly slime on your fingers. But slow the squirt down and the majesty of the velvet worm glue gun becomes clear. Each of the cannons is oscillating, waving horizontally to create a stream that looks like a sine wave. Yet that isn't a muscular action. It's simply a trick of physics: Just as a garden hose left to its own devices at full blast will flop around, so, too, do the velvet worm's guns, as many as sixty times a second. And so the two streams will cross over each other, creating more of a shotgun blast than a precise shot, to hit prey as far as eight inches away.

Once the glue coats the victim, it immediately coagulates. The prey is doomed, and struggling only helps ensnare it more. Never one to be rushed, the velvet worm approaches slowly. It crawls up on top of the victim and searches for a good spot to drive its blades into—maybe the less-armored joint of a cricket would be nice—and injects a saliva that begins breaking down the thing's insides. At its leisure, the worm gobbles up the glue that's lying around going to waste, sucks the juices out of the prey, eats the odd limb here and there, and then shuffles on.

Perhaps the most enthralling and methodical (in a good way) experiments with such velvet worm hunts come from two biologists who set up shop in Trinidad in the 1980s. They worked with a species that prefers to get nice and close to its victims, and they recorded its tactics in great detail. The velvet worm would begin by sneaking up on the prey and gently probing it with the antennae, often quite thoroughly, and somehow without raising suspicion. When it was satisfied with its choice, it would blast its goo at the victim, so rapidly that the "emission of glue was undetectable, but the prey would suddenly appear covered in a network of beaded threads of glue." Usually one blast sufficed, but if the prey struggled, the worm would hit it with up to thirty additional blasts, many of them aimed at the limbs, and in the case of spiders, at the fangs. Once the prey was subdued, the velvet worm injected its saliva and spent as long as an hour consuming the threads around the corpse. Finally turning its attention to the meal itself, the hunter took perhaps another ten hours eating the quarry.

About that glue. The researchers found that the stored glue made up an average of 11 percent of the velvet worm's body mass. An impressive figure, but those reserves can quickly run out, especially when the velvet worm is tackling bigger targets. And that's a

SPIT-TAKE

Sharpshooting isn't the territory of just terrestrial hunters. Archerfish target insects crawling along trees above flooded mangrove forests, spitting high-speed jets of water to knock them off their perches. That's no small feat, considering the fish has to correct both for the fact that light bends when it hits water, making the insect appear where it isn't, and that gravity tugs on the jet, making it more of an arch than a straight shot.

More impressive still, the jet speeds up as it approaches the target, which would seem to, well, defy the laws of physics. The trick here is that the archerfish spits so the rear of the stream is traveling faster than the front. The jet turns into blobs, and when the faster blobs at the back reach those at the front, they form a larger blob that's traveling at a higher speed. Really, it's more like firing a rocket than an arrow, so I guess we should call them rocketfish instead of archerfish. Wouldn't want it to go to their heads, though.

dangerous prospect when you consider that it takes weeks for the worm to fully replenish the reserves, during which time it may not be able to hunt as effectively or defend itself from its own enemies, which it aims to slow down with glue strikes. What the researchers found, though, is that the velvet worm tailors the amount of glue it fires based on the size of its prey, using as much as 80 percent of its reserves to tackle its biggest prey, like large crickets.

Think of it as an investment. It seems counterintuitive, but going after small prey is a risk for the velvet worm. If it ends up blowing through too much glue trapping the tiny thing, that's a crummy investment, because the energy it gains from eating the prey can't offset the lost material, which requires energy to produce. Accordingly, in the Trinidad experiments velvet worms preferred larger prey over their smaller counterparts. Such a big meal justified the expense of glue. But the hunter can't risk going *too* big. The largest prey could well escape, making off with the glue the worm would normally consume and recycle. So if food is plentiful, the predator aims for victims somewhere in the middle, not too small to be a waste of glue, and not too big to abscond with that precious resource.

It's easy to forget that out there in nature choices like these are a matter of life and death. I don't imagine you've had to make the decision recently whether or not walking to the corner store would use up more energy than you'd get from the six-pack you'd bring home (forgive me, I'm nearing my deadline, and am familiar with such pursuits), but nature scoffs at a waste of energy. It's why alligators laze around so much, and why being a glacially slow snail makes sense. Plants aren't about to get up and flee from you, so there's no sense in being speedy and burning all that energy.

It makes sense, that is, unless you're a snail that craves meat. Then you're going to need one hell of a solution to get by.

Geography Cone Snail

PROBLEM: Snails also aren't celebrated for their speed.

SOLUTION: To capture fish, the cone snail drugs them first by releasing chemicals into the water, before strolling up and enveloping the prey with its balloonlike mouth.

Out on an island in the Great Barrier Reef in the year 1935, a man picked up a *Conus geographus*, or geography cone snail. Holding it in his palm, he proceeded to scrape off the thin cuticle covering the shell. This, however, did not suit the snail, which fired its venomous harpoon into the man's palm. And so began the numbness. Ten minutes later, the man's lips had gone stiff, and after another ten minutes his vision went double. A half hour after the sting, his legs were paralyzed, before he slipped into a coma. In four more hours, he was dead.

Strangely enough, just three years prior to the incident, another man, this one on an island off of Africa, made the same mistake, holding a snail and scraping it with his knife. Following the inevitable sting, his entire body went into paralysis. Luckily someone got him to a doctor, who treated the victim and recommended "that he should be given general massage in addition." I'm not a betting man, but I'm reasonably confident the massage didn't play much of a part in his eventual recovery. But then again, it probably didn't hurt either.

You may already be familiar with cone snails. They are, after all, some of the most venomous creatures on Earth, propelling toxic harpoons into fish so quickly that the strike is all but invisible to the naked eye. But the geography cone snail belongs to a relatively obscure group among them called the net hunters. They don't just approach a fish and fire a venomous harpoon into it. They go for entire *schools* of fish, and they take their sweet time with it, sedating the victims with a cloud of toxins and strolling in for the kill. Because while other hunters have the luxury of speed, the net-hunting snails have hungry mouths attached to sluggish bodies. Their solution to holding on to their prey isn't brawn, but a brutal, sophisticated mix of both chemical and ballistic warfare.

Something weird happens when a geography cone snail approaches a school of fish. Well, a couple weird things happen. First, the fish don't seem to give a hoot. They're aloof, lolling there wide-eyed. They don't even seem to give a hoot when the snail expands its huge mouth like a fleshy hot air balloon. Slowly the mouth envelops several clueless fish, until the snail at last fires a harpoon into each victim, paralyzing it almost instantly. The mouth balloon shrinks, and the fish go quietly into death.

Which would seem like a silly thing to be so relaxed about. But it turns out that the fish are under a kind of spell: The geography cone snail has evolved to deploy insulin as a chemical weapon. By releasing massive amounts of the hormone into the water, the snail makes its prey's blood sugar levels plummet. Hypoglycemic shock sets in. The fish grow sluggish and confused as their nervous systems glitch. In effect, they're hypnotized, conscious yet locked up, like you or me suffering a fit of sleep paralysis. But what's even weirder is that insulin in an invertebrate like a geography cone snail and insulin in a vertebrate like a fish are chemically distinct. Yet in addition to producing invertebrate insulin for its own body, the snail has evolved fishlike insulin that it has weaponized to overload its prey.

But there are still more toxins at work here to sedate the snail's prey, forming a complex mixture that biologist Baldomero Olivera, an expert on cone snail venom, calls the nirvana cabal. ("Nirvana" is pretty self-explanatory, and "cabal" because, according to Olivera, the various substances in the mixture work together to overwhelm the fish, just as a cabal of humans would do to a government.) The geography cone snail has additional peptides that further sedate the fish, quieting their nervous systems. In fact, researchers have used one of these peptides in clinical trials to treat epilepsy, which stems from abnormal levels of electrical activity in the brain.

So the geography cone snail has hypnotized and drawn several fish into its mouth, but it has to pull one last trigger. Attached to its venom gland is a cache of around twenty harpoons, which are in fact modified teeth. It'll fire one into a dazed fish, load another harpoon, and fire that into the next, right on down the line. It's a massacre. Dosed with enough venom, known as a motor cabal, to kill an adult human, the fish's neurons short-circuit as their muscles shut down. Death probably comes from asphyxiation, since fish have to be able to pump water through their gills in order to breathe.

Now, the geography cone snail's more well-known cousins, the so-called hook-and-line cone snails, hunt in a different way by harpooning just a single fish, and therefore they deploy different toxins. In this group, when the harpoon spears the prey, the first toxins that go to work are a group aptly named the lightning cabal (there are lots and lots of chemicals working in concert here, in some species as many as two hundred, and each species has its own unique mix). This messes with a fish's neurons like the motor cabal does, but instead of short-circuiting them, it overloads them so they fire constantly. The fish is essentially Tasered, seizing up and twitching uncontrollably. The motor cabal takes longer to kick in, thus the lightning immobilizes the fish so the snail can reel it in, followed by the delayed effects of the second cabal, which put the prey down for good. Yet a net hunter like the geography cone snail has no need for that regimen. By pre-sedating schools of fish before it even makes contact, the snail can envelop them without a fight. Plus, if the snail hits the prey with the lightning cabal at the harpooning, it'd have a spazzing school of fish in its mouth, potentially leading to injury. It has them right where it wants them, no Tasering necessary.

It all adds up to a dizzyingly sophisticated and powerful weapon, evolved over generation after generation. So remember, no scraping snails with pocketknives. Unless you were looking for a doctor-mandated massage—then be my guest.

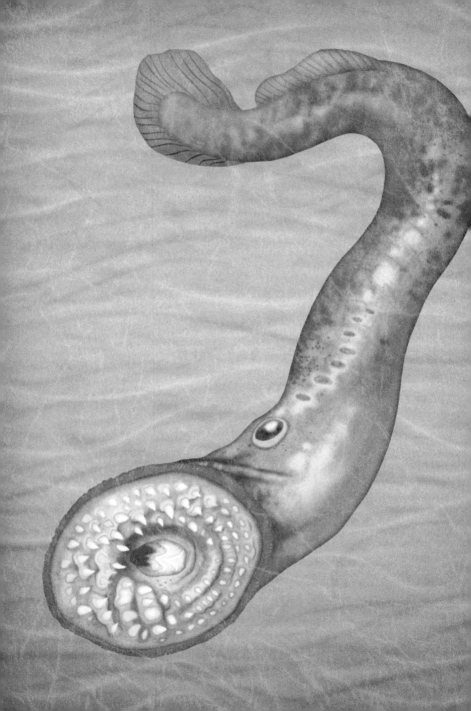

Lamprey

PROBLEM: Fish don't take too kindly to traumatic parasitization.

SOLUTION: The lamprey has evolved to look like the desert pit monster that ate Boba Fett in *Star Wars*: It suctions on to its thrashing victim with a disklike mouth packed with dozens of hooked teeth and will often drill right down to the bone.

Raining frogs is one thing, but raining toothy tubes of flesh is a different anomaly entirely. It was June 2015, and the Alaska Department of Fish and Game was getting "calls about arctic lamprey found in strange locations" around Fairbanks. Someone had discovered the eel-like fish, with its fearsome suction-cup mouth it uses to suck the blood out of other fish, in a parking lot—hardly its natural habitat. Another lamprey materialized on a lawn. And two others around town. Yet there had been no reports of waterspouts, that classic generator of raining aquatic beings. But the Fish and Game folks had a solid clue. V-shaped bruises ran along the lampreys' bodies: gull bites. Maybe there happened to be a lot of lampreys in a nearby river at the time. And maybe gulls like lampreys. And maybe sometimes gulls drop lampreys in midair. And thus does it appear to rain bloodsucking fish on the Golden Heart City of Fairbanks, Alaska.

On the other side of North America, in the Great Lakes, the lamprey isn't a mere parking-lot oddity. Here, it's an invasive menace that has devastated populations of native fish: At one point in the twentieth century, the lamprey crashed the fishery in the

Great Lakes to 2 percent of its normal catch. These days, eradication efforts have brought the invasive beast under some measure of control, but no realistic human being is expecting to rout the lamprey entirely. With so many tributaries connected to the lakes, cash-strapped officials can hope only for suppression, not outright defeat.

This story begins 360 million years ago with the earliest known lamprey fossil, which looked a lot like modern varieties. That number is, simply put, ridiculous. Life complex enough to be recognizable to us as life has been around for 570 million years, while the first mammals showed up around 200 million years ago. Yet the lamprey has survived largely unchanged, through all kinds of mass extinctions, for almost 400 million years. (As it happens, the lamprey's only close relative is our friend the hagfish, a similarly ancient species. And indeed the two share a lot of characteristics, including a skeleton of cartilage instead of bone.)

WHO ARE YOU CALLING A FOSSIL?

You'll hear the phrase "living fossil" tossed at the lamprey and any number of other ancient species, but that language is problematic. It implies that the 360-million-year-old lamprey fossil is identical to the lampreys that swim Earth's waters today, and that's an impossibility. The lampreys didn't survive for this long, over so many catastrophic events and climate changes and introductions of new predators and extinctions of potential prey, by being immutable. "Living fossils" may look like they've gone unchanged, but subtle variations over evolutionary time have allowed them to roll with the punches, as it were—if you could consider something like an asteroid hitting your planet right in the face as a "punch."

Some of these species attach to their host and sip a bit of blood, while the flesh eaters among them go further, gouging out plugs of meat and leaving their victims with gaping wounds, and likely death by blood loss or infection. For their exploits both types of lampreys are equipped with a mouth known as an oral disk, which is loaded with dozens upon dozens of hooked, fanglike teeth that help their owner get a purchase on a victim (aided by special structures along the disk's edge that form a suction). Like a shark's teeth, these are replaceable—*highly* replaceable: The vampiric species that's invaded the Great Lakes, the sea lamprey, may replace its teeth as many as thirty times in two years.

What sets the bloodsuckers and flesh eaters apart, though, is something called a piston, which fires in and out of the center of the oral disk. A piston is like a tongue, only made out of teeth: one tooth that flicks vertically and two others that flick horizontally, creating a complex rasping mechanism. In bloodsuckers, that vertically firing tooth in the piston is shaped like a W to scrape away flesh and draw blood—these species also secrete an anticoagulant to ensure that blood flows freely—while in flesh eaters the tooth is more of a U shape that gouges out chunks of flesh (think of the former shape as sandpaper and the latter as a chisel). The whole mess is astonishingly effective. In the Great Lakes a single sea lamprey can kill forty pounds of fish in a year. At times, only one in seven fish here will survive a lamprey attack, and the fishery is suffering for it.

Really, though, it's our own fault. You may be wondering by now what business a sea lamprey would have conquering freshwater lakes. But they have every right to be here. Sea lampreys begin their lives in freshwater streams, biding their time as harmless larvae for up to seven years. During that stage, their mouths are just squishy holes, with which they grab planktonic creatures floating by. When they mature, they head out to sea to parasitize fish, but return to the streams to spawn (interestingly, the larvae guide adults back from the sea to optimal spawning grounds by releasing pheromones). Thus, capable of living in freshwater, the

THE LAMPREY-LOVING HENRY AND THE LAMPREY-LOVING HENRY WHO CAME BEFORE HIM

The lamprey's mouth, not to mention its habits, hasn't seemed to bother European nobility in the slightest. King Henry I, visiting Normandy in 1135, is rumored to have eaten so many lampreys at a feast that he up and died. Henry V, apparently unconcerned about the fate of the Lamprey-Loving Henry That Came Before Him, also demanded lamprey while visiting Normandy.

Even in 2012 for Queen Elizabeth II's diamond jubilee, the city of Gloucester presented a traditional regal lamprey pie. Only those weren't British lampreys, for the fish are scarce over there in these modern times. But not here in the States. So a representative of the Great Lakes Fishery Commission flew across the pond and presented Gloucester with sacrificial American lampreys. Thus did the queen get her pie while the United States got rid of a few lampreys. Ah, diplomacy.

parasitic adult sea lamprey jumped at the opportunity to invade the lakes when in the early twentieth century engineers deepened a canal that bypassed Niagara Falls, the natural barrier keeping the lampreys out. The floodgates opened, and the Great Lakes haven't been the same since.

This is the sad nature of the invasive species. The Great Lakes ecosystem had been evolving for millennia without having to worry about the sea lamprey, and when the beast showed up, the complex web of organisms went into shock. The problem had arrived so suddenly there wasn't time for the victims to find a solution, for an ecosystem is a delicate dance of checks and balances: A predator evolves a weapon and the prey evolves a defense in a graceful game of one-upmanship. Sure, it's inevitable over evolutionary time that a new species may arrive in an ecosystem without human assistance—a bird riding a hurricane to an island, for instance—but our global economy has helped shuttle innumerable species around the world, be it a mussel stowed away in the ballast of a ship or a seed caught in a boot. The sea lamprey simply took advantage of the freebie we threw at it.

Canadian and American wildlife officials around the Great Lakes are waging perpetual war against the lamprey. At their disposal are a variety of weapons. There's a poison, called lampricide, that targets the things without harming other fish. Barriers that block their upstream migrations also help. But no method is more creative than the good ol' lamprey mass sterilization. Crews will round up a whole mess of lampreys, chemically castrate the males, and ship the females off for research. Infertile males released back into the lakes will compete for the right to mate, yet will fire blanks when it's time to perform. It sounds far-fetched, but it's helping keep the population in check. And hell, it never hurts to emasculate a few lampreys.

Assassin Bug

PROBLEM: Some insects just can't be bothered to get gnawed on alive.

SOLUTION: The assassin bug impales its prey with super-elongated, needle-like mouthparts, sucks out their juices while they're attached to its face, and then sticks their corpses to its back as camouflage.

Perhaps evolution's most beautiful dichotomy is that it is at once a force of total chance and total order. The way your parents' genes combined to make you and your siblings is random. But the traits you bear make you either more suitable or less suitable to the forces of natural selection—for very specific, nonrandom reasons. A food chain may seem like chaos, yet there's an order to things: A mountain lion will maul a rabbit, and never the other way around. But a mountain lion is a mountain lion and a rabbit is a rabbit because of the random mutations that built their species, and indeed every species on Earth. Thus from randomness arises order.

So we might call it an inevitability that in the course of his travels around South America Darwin would fall prey not to a predatory cat, but to a far more common foe: the assassin bug, with its huge, strawlike mouthparts that it drives into flesh and uses to suck out blood. "At night I experienced an attack (for it deserves no less a name) of the *Benchuca*, a species of Reduvius, the great black bug of the Pampas," Darwin writes in *The Voyage of the Beagle*. "It is most disgusting to feel soft wingless insects, about an inch long, crawling over one's body. Before sucking they are quite thin, but afterward they become round and bloated with blood, and in this state are easily crushed."

Darwin continues, describing a peculiar scene that reads more

"I AM AFRAID THE SHIP'S ON FIRE. COME AND SEE WHAT YOU THINK OF IT."

It's worth rehashing the radically different travels of Darwin and Wallace, on account of them being so, well, radically different. Assassin bugs tormented Darwin, sure, but one of Wallace's ships caught on fire. Like, the whole thing. And then it sank.

Sitting in his cabin on his way back to London from a collecting expedition in South America, a fever-wracked Wallace found the captain's head peeking through his door, delivering what might be the greatest understatement in the history of natural history: "I am afraid the ship's on fire. Come and see what you think of it." The ship was, in fact, a good amount on fire. Wallace made his way to the boats, rescuing only a few notes and sketches—and consigning case after case of specimens to the flames. Sliding down the rope to the escape boat, Wallace burned the skin off his palms, and spent the next ten days having the sun bake the flesh off of him, with no room to curl up and have a sleep, until a vessel bound for England happened to cross the path of the sad crew. Even then, on the voyage back home they damn near starved as nasty storms split their sails. Finally reaching the motherland, Wallace made up for lost food. "Here we are on shore . . . such a dinner! Oh!" he wrote in a letter. "Beef steaks and damson tart, a paradise for hungry sinners." Sinners? Good God, man, go easy on yourself.

like science fiction than natural history. He caught an assassin bug in Chile—one with an empty belly—and introduced it to a few of his comrades. "When placed on a table, and though surrounded by people, if a finger was presented, the bold insect would immediately protrude its sucker, make a charge, and if allowed, draw blood." The attack wasn't painful, Darwin notes, and one officer at the table let the thing feed on him for ten minutes, during which time the assassin bug "changed from being as flat as a wafer to a globular form."

Really, there are few insects as clever, ferocious, or unnerving as the aptly named assassin bug, which is pretty much a giant mosquito. While the bolas spider lies in wait, and the tiger beetle runs down its prey, the seven thousand species of assassin bug take an altogether more ninjalike approach. They're sneakers. Ambushers. They have all kinds of cunning tactics, but what remains constant among them is their weapon: an elongated mouth, called a rostrum, which they slam into their victims. Once through the skin or exoskeleton of the prey—be it a mammal or a fellow insect, depending on the species of assassin—a sheath pulls back and the actual mouthparts emerge. At this point the assassin releases a toxin that paralyzes the victim, which is still very much alive as its insides start to liquefy. The killer then drinks up the soup from the prey impaled right there on its face.

The most bizarre hunting tactic among the many bizarre hunting tactics of the assassin bugs is what I like to call the Backpack of My Enemies. After draining the insides of prey like ants, some species will arrange the corpses on their sticky backs, piling the bodies so high that the Backpack of My Enemies ends up growing far taller than the assassin itself, making the hunter look like a macabre Santa with a sack of the worst presents a kid could ever ask for. That, as you can imagine, makes the assassin unpalatable to its own enemies, plus has the added benefit of bestowing on it the scent of its prey, thus allowing the hunter to better sneak up on its victims' comrades.

Another variety goes one step further with the masquerade.

TERRIFIED OF SPIDERS?
TERRIFIED OF SNAKES?
THEN DO I HAVE A CRITTER FOR YOU

The feather-legged assassin bug gets points for gutsiness, but the equally hyphenated spider-tailed horned viper of Iran wins big-time for the creepiest lure in nature. Think of a rattle-snake whose tail doesn't rattle, but instead looks just like a spider, with a bulb for an abdomen and little offshoots that look like legs. The well-camouflaged snake curls up and gives that tail a wag for hungry birds in the vicinity. When one takes the bait, the snake strikes, loading the greedy bugger with venom. It isn't quite as fashionable as leg bristles, but then again, *it's a spider-snake*.

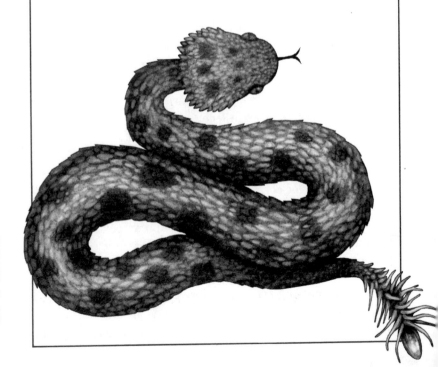

Termites of course have mounds, which emit lots of nice termite smells. So instead of coating itself in corpses, the assassin that targets these termites applies the nest material to its back to become one with the environment. After spearing a termite and sucking it dry, the assassin will keep the victim on the rostrum and dangle it into the mound. This is of some concern to the victim's compatriots not because of a sort of emotional attachment, but because as social insects, termites react to sickness and death like the ants under attack from zombifying fungi or decapitating flies do. They're programmed to remove any dead or dying termites from the mound, a behavior the assassin is there to exploit. Any rescuers are themselves stabbed, drained, and dangled, one after the other, perhaps dozens in a sitting.

Still another species doesn't just lure ants, but waits for the ants to attack it, a risky and extreme rarity in nature (lures like the anglerfish's are one thing, but the hunter doesn't want to die in the process). The nymph feather-legged assassin bug—with its big, fluffy masses of bristles on its rear pair of limbs—positions itself among foraging ants and rapidly twitches its legs. When the dupe comes to investigate, the assassin doesn't react until the much larger ant clamps onto its leg and starts dragging it away. But sure enough, the assassin whips around onto the ant's back and impales it.

The assassin bugs that attack humans, known as kissing bugs because they're attracted to exhalations and are therefore prone to stab around the mouth, aren't nearly so clever, but they're still deadly. It isn't the wound itself that kills, but the feces the assassin leaves behind. If the victim gives the wound a scratch, the surrounding droppings can make their way inside, transmitting a protozoan parasite that causes Chagas disease: severe cardiac and digestive problems that may only manifest decades after a bite.

It's this disease, and the South American assassin bugs it rode in on, that some scholars have fingered as the cause of Darwin's death. After all, he suffered from chronic health problems—fatigue, vomiting, abdominal pain—that all but incapacitated him, particularly in his later years. Others say it was Crohn's disease, or even lactose intolerance, or something called cyclical vomiting syndrome, or any number of other maladies. The matter is, for some strange reason, still a matter of fierce debate.

What's certain, though, is that the ultimate predictability of nature, death, came on April 19, 1882, to one of the most brilliant random assemblages of human genes ever to walk the Earth. The allies who had defended Darwin's civilization-shaking ideas lobbied to have him buried at Westminster Abbey, a seat of the religious establishment the man had so rankled. And after some wrangling, so it was.

Through a congregation of the titans of science and government and faith, ten pallbearers marched with the coffin. Among them was Darwin's cherished friend Alfred Russel Wallace, lanky as ever. Wallace and the expired mind he carried had discovered the greatest idea in human history, an idea so radical that it sent religious figures into a frenzy, yet was so undeniably true that here the pious men were, accepting Darwin's remains. Darwin and the evolutionists had brought chaos to the idyllic Victorian conception of nature, insisting that survival itself is just one big problem. No creature can avoid its eventual demise, but these bodies of ours are pretty good shots at a solution, however vulnerable they may be to assassin bugs, lactose intolerance, or what have you. Sure, we don't have toxic hairs to protect us, we can't invade the bodies of other creatures, and we can't release clouds of goo to choke our enemies. But damn it, at least we've got our friends.

A Few Parting Words

L ife, as you may have gathered from this book, is death. Destruction comes from above, from below, sideways and diagonally, from parasites and predators and the environment itself. For billions of years life has flourished on this planet, and almost every single bit of that life is today buried in the ground. But that's not for lack of trying. Creatures have gone to great lengths to survive and reach the ultimate goal: make babies. They've faced their trials and found their solutions, however accidentally, evolving and evolving and evolving to culminate in the incredible array of life with which we now have the honor of sharing a planet.

It shocks the human mind that a creature could evolve to weaponize insulin or snot, to mind-control an ant or caterpillar, to fight with its penis or its mustache. And the human mind stretches to its limits to comprehend that all of this is automatic. *Boom*, that first microorganism came into being, and the rest has since fallen into place. No guiding hand, no purpose, just species battling each other and the elements. They create problems for the opposite sex or for their prey, problems that the opposite sex or their prey in turn solve. Species grow faster or tougher, taller or tinier, marching through time, adapting to one trial after another.

It's easy to forget this isn't over. Species will keep on evolving, sometimes right before our very eyes. There's no way to predict if the aye-aye's finger will grow still longer, or if antechinus sex will grow still more frenzied, but we can say with confidence that evolution now must grapple with a complicating factor: us.

Earth has never seen a force like humanity. We pull down mountains and gouge great holes in the earth. We've transformed our climate and polluted the bejesus out of the seas. We've driven countless species to extinction as others struggle to hold on. Humanity is Earth's biggest problem, and it'll be the species that stumble upon solutions to us that will make it. Individuals that can better withstand warming and acidifying seas, for instance, will survive to pass down their genes for this resilience.

Not since the pioneering days of Maria Sibylla Merian and Charles Darwin and Alfred Russel Wallace has the work of the natural historian been more critical. Saving the species we're imperiling requires understanding them. And so Brian Fisher, iPad in hand, slogs through the jungle in search of the hero ant. Tierney Thys wrangles the ocean sunfish, the biggest bony fish in the world, tagging them so she can study their movements in rapidly changing oceans. And Mariella Superina, after all these years, still scours the deserts of Argentina in search of the mysterious pink fairy armadillo. They're just three explorers among the multitudes of scientists trying to make sense of this mad planet of ours.

Who knows—maybe we'll end up solving the problem that is us. And I don't mean extincting ourselves. Evolution gave us these great big brains, after all. Perhaps we can figure out how to live on this planet without destroying it. Because if we don't, the only other option is space.

But hell, at least we could take a few water bears with us for company. Or is that wildly irresponsible? That's wildly irresponsible, isn't it.

Acknowledgments

My grandparents lived up in the mountains, far away from the suburbs, where things called animals lived. For instance, the bats, my grandfather explained to a tiny me, could squeeze right between the backboard of his basketball hoop and the garage it was flush against. *But there's just no clearance*, I remember protesting, though clearly not in that exact language. Yet true enough they roosted there, no doubt scoffing at my judgment, as bats are wont to do.

Thank you, Grandpa, and thank you, Grandma, for bringing me into that world. Thank you, Dad, for rescuing the tick-infested pet rabbit we found in a park, a sad creature I named Bugsy, because I was an asshole. Thank you, Mom, for the snark, which either got readers through this book or made them put it down after the first chapter. For the former, thanks for sticking with it. Thank you, Melissa, for giving me permission to write about your kid pooping on a lawn, thus entering his exploits into the public record.

A thousand thanks to Superagent David Fugate. I seriously have no clue how you have the energy for all of this. You're a master. And thank you to my editor over there at Penguin, Meg Leder, for guiding this book to a place where it made sense other than in my own head. You're a fantastic editor. You as well, Shannon Kelly. Thank you for laughing at my dumb jokes.

Brian Chen, you know what you've done.

Sorry, that wasn't meant to sound like you stabbed me once in a bar fight. I mean I really appreciate what you've done.

My respect to the entire *Wired* clan. For Betsy Mason taking the chance on a half-baked—perhaps quarter-baked—idea for a column. For Chuck Squatriglia and his many tasteful edits, and less than tasteful taste in metal.

Then there's the trio of brilliant scientific minds who were so kind as to give the manuscript a gander. Danielle Venton, radio personality and friend of goats. Nadia Drake, jungle traipser and ofttimes horse. And Gwen Pearson, one hell of an entomologist who knows better than to bother with mammals like goats and horses.

I'd like to thank every single biologist who ever made the mistake of answering my phone call or e-mail asking for an interview. Some of you have been amazing, some of you have been rotten bastards, but every single one of you has put on a clinic for me. I'm humbled, immensely grateful, and a better person because of you all.

And thanks to all the creatures. I know you can't read this, but I'd feel lousy not saying it.

Bibliography

CHAPTER 1:
You Absolutely Must Get Laid

Antechinus

Darwin, C. (2009). *On the Origin of Species: A Facsimile of the First Edition.* Cambridge, MA: Harvard University Press.

Fisher, D., C. Dickman, M. Jones, and S. Blomberg. (2013). Sperm Competition Drives the Evolution of Suicidal Reproduction in Mammals. *PNAS*, 10.1073/pnas.1310691110.

Kemper, S. (2008). Who's Laughing Now? *Smithsonian.* Retrieved from http://www.smithsonianmag.com/science-nature/whos-laughing-now-38529396/?all&no-ist.

Simon, M. (2014). Absurd Creature of the Week: This Marsupial Has Marathon Sex Until It Goes Blind and Drops Dead. *Wired.* Retrieved from http://www.wired.com/2014/05/absurd-creature-of-the-week-this-marsupial-has-marathon-sex-until-it-goes-blind-and-drops-dead/.

———. (2014). Fantastically Wrong: The Poor, Misunderstood Hyena Can't Help That It Has Weird Sex. *Wired.* Retrieved from http://www.wired.com/2014/05/fantastically-wrong-sexually-deviant-hyenas/.

Tyson, P. (2007). Evolution Down Under. *PBS.* Retrieved from http://www.pbs.org/wgbh/nova/evolution/evolution-down-under.html.

Yong, E. (2013). Why a Little Mammal Has So Much Sex That It Disintegrates. *National Geographic.* Retrieved from http://phenomena.nationalgeographic.com/2013/10/07/why-a-little-mammal-has-so-much-sex-that-it-disintegrates/.

Anglerfish

Bioluminescence Questions and Answers. (2015). Latz Lab, UC San Diego. Retrieved from https://scripps.ucsd.edu/labs/mlatz/bioluminescence/bioluminescence-questions-and-answers/.

Fairbairn, D. (2013). *Odd Couples: Extraordinary Differences Between the Sexes in the Animal Kingdom.* Princeton, NJ: Princeton University Press.

Gould, S. (1994). *Hen's Teeth and Horse's Toes: Further Reflections in Natural History.* New York: W. W. Norton.

Johnsen, S. (2012). *The Optics of Life: A Biologist's Guide to Light in Nature.* Princeton, NJ: Princeton University Press.

Pietsch, T. (1975). Precocious Sexual Parasitism in the Deep Sea Ceratioid Anglerfish, *Cryptopsaras Couesi Gill. Nature*, 10.1038/256038a0.

———. (2005). Dimorphism, Parasitism, and Sex Revisited: Modes of Reproduction Among Deep-Sea Ceratioid Anglerfishes. *Ichthyological Research*, 10.1007/s10228-005-0286-2.

———. (2009). *Oceanic Anglerfishes: Extraordinary Diversity in the Deep Sea*. Berkeley and Los Angeles: University of California Press.

Simon, M. (2013). Absurd Creature of the Week: The Anglerfish and the Absolute Worst Sex on Earth. *Wired*. Retrieved from http://www.wired.com/2013/11/absurd-creature-of-the-week-anglerfish/.

Yoshizawa, K., R. Ferreira, Y. Kamimura, and C. Lienhard. (2014). Female Penis, Male Vagina, and Their Correlated Evolution in a Cave Insect. *Current Biology*, 10.1016/j.cub.2014.03.022.

Flatworm

Bootlace Worm. (n.d.). NAFC Marine Centre. Retrieved from http://www.nafc.uhi.ac.uk/departments/marine-science-and-technology/discovery-zone/discovery-zone-map/bootlace-worm.

Brennan, P. (2013). Why I Study Duck Genitalia. *Slate*. Retrieved from http://www.slate.com/articles/health_and_science/science/2013/04/duck_penis_controversy_nsf_is_right_to_fund_basic_research_that_conservatives.html.

Michiels, N., and L. Newman. (1998). Sex and Violence in Hermaphrodites. *Nature*, 10.1038/35527.

Milius, S. (2009). Hermaphrodites Duel for Manhood. *Science News*, 10.2307/4010187.

Natural Selection: Charles Darwin and Alfred Russel Wallace. (n.d.). University of California, Berkeley. Retrieved from http://evolution.berkeley.edu/evolibrary/article/history_14.

Newman, L., and L. Cannon. (2003) *Marine Flatworms: The World of Polyclads*. Clayton South: CSIRO Publishing.

Pennisi, E. (2011). Immune System Protects Female Bedbugs from Traumatic Sex. *Science*. Retrieved from http://news.sciencemag.org/evolution/2011/08/immune-system-protects-female-bedbugs-traumatic-sex.

Ramm, S., A. Schlatter, M. Poirier, and L. Schärer. (2015). Hypodermic Self-Insemination as a Reproductive Assurance Strategy. *Proceedings of the Royal Society B*, 10.1098/rspb.2015.0660.

Roughgarden, J. (2010). Beauty and the Beast. *American Scientist*. Retrieved from http://www.americanscientist.org/bookshelf/pub/beauty-and-the-beast.

Stutt, A., and M. Siva-Jothy. (2001). Traumatic Insemination and Sexual Conflict in the Bed Bug *Cimex Lectularius*. *PNAS*, 10.1073/pnas.101440698.

Mustache Toad

Darwin, C. (1860). Letter to Asa Gray. Retrieved from http://www.darwinproject.ac.uk/letter/entry-2743.

Emlen, D. (2014). *Animal Weapons*. New York: Henry Holt.

Hudson, C., and J. Fu. (2013). Male-Biased Sexual Size Dimorphism, Resource Defense Polygyny, and Multiple Paternity in the Emei Moustache Toad (*Leptobrachium Boringii*). *PLOS One*, 10.1371/journal.pone.0067502.

Izzo, T., D. Rodrigues, M. Menin, A. Lima, and W. Magnusson. (2012). Functional Necrophilia: A Profitable Anuran Reproductive Strategy? *Journal of Natural History*, 10.1080/00222933.2012.724720.

Simon, M. (2014). Absurd Creature of the Week: This Toad Grows a Spiky Mustache and Stabs Rivals for the Ladies. *Wired*. Retrieved from http://www.wired.com/2014/02/absurd-creature-week-toad-grows-spiky-mustache-stabs-rivals-ladies/.

Toadfish

Brahic, C. (2008). "Horror Frog" Breaks Own Bones to Produce Claws. *New Scientist*. Retrieved from http://www.newscientist.com/article/dn13991-horror-frog-breaks-own-bones-to-produce-claws.html#.VN_G1nbWTVo.

Brantley, R., and A. Bass. (2010). Alternative Male Spawning Tactics and Acoustic Signals in the Plainfin Midshipman Fish *Porichthys notatus* Girard (Teleostei, Batrachoididae). *Ethology*, 0.1111/j.1439-0310.1994.tb01011.x.

Carson, R. (1989). *The Sea Around Us*. New York: Oxford University Press.

Do Lovesick Fish Sing in Sausalito? [Editorial] (August 12, 1985). *Marin Independent Journal*, A10.

Dorcas, M., and W. Gibbons. (2011). *Frogs: The Animal Answer Guide*. Baltimore: Johns Hopkins University Press.

Luxury Condoms: Ultrasensitive Thanks to Fish Bladders. (2015). Museum of Contraception and Abortion Retrieved from http://en.muvs.org/contraception/condoms/fischblasenkondom-auf-holzpenis-id1995/.

McCosker, J. (1986). The Sausalito Hum. *Journal of the Acoustical Society of America* 80(6): 1853–54.

———. (June 1986). In Sum, It Was Some Hum. *Discover Magazine*.

CHAPTER 2:
You Can't Find a Babysitter

Ant-Decapitating Fly

Areawide Fire Ant Suppression. (2015). USDA Agricultural Research Service. Retrieved from http://www.ars.usda.gov/sites/fireants/Imported.htm.

Geist, L. (2012). Fire Ants Mobilize in the Aftermath of Disaster. University of Missouri. Retrieved from http://extension.missouri.edu/news/DisplayStory.aspx?N=1596.

Hölldobler, B., and E. O. Wilson. (1995). *Journey to the Ants*. Cambridge, MA: Belknap Press.

Red Imported Fire Ant (*Solenopsis Invicta*). (n.d.). Desert Museum. Retrieved from http://www.desertmuseum.org/invaders/invaders_fireant.php.

Rhatigan, J. (2010). *Book of Science Stuff*. Watertown, MA: Imagine Publishing.

Simon, M. (2013). Absurd Creature of the Week: This Fly Hijacks an Ant's Brain—Then Pops Its Head Off. *Wired*. Retrieved from http://www.wired.com/2013/12/absurd-creature-of-the-week-this-fly-burrows-into-an-ants-brain-then-pops-its-head-off/.

Wallace, A. R. (2000). *The Malay Archipelago*. North Clarendon, VT: Periplus Editions.

Glyptapanteles Wasp

Grosman, A., et al. (2008). Parasitoid Increases Survival of Its Pupae by Inducing Hosts to Fight Predators. *PLOS One*, 10.1371/journal.pone.0002276.

Simon, M. (2013). Absurd Creature of the Week: Burrowing Botfly Grows Huge Feasting on Your Flesh. *Wired*. Retrieved from http://www.wired.com/2013/10/absurd-creature-of-the-week-botfly/.

———. (2014). Absurd Creature of the Week: The Wasp That Lays Eggs Inside Caterpillars and Turns Them Into Slaves. *Wired*. Retrieved from http://www.wired.com/2014/10/absurd-creature-week-glyptapanteles-wasp-caterpillar-bodyguard/.

Whitfield, J. (2004). Ichneumonoidea. Tree of Life Project. Retrieved from http://tolweb.org/Ichneumonoidea.

Asp Caterpillar

Bishopp, F. (1923). The Puss Caterpillar and the Effects of Its Sting on Man. *United States Department of Agriculture Department Circular*, 10.5962/bhl.title.65870.

Diaz, J. (2005). The Evolving Global Epidemiology, Syndromic Classification, Management, and Prevention of Caterpillar Envenoming. *American Journal of Tropical Medicine and Hygiene* 72(3): 347–57.

Eagleman, D. (2007). Envenomation by the Asp Caterpillar (*Megalopyge Opercularis*). *Clinical Toxicology*, 10.1080/15563650701227729.

Hall, D. (2012). Featured Creatures: Puss Caterpillar. University of Florida. Retrieved from http://entnemdept.ufl.edu/Creatures/MISC/MOTHS/puss.htm.

Heard, K. (2016). *Maria Merian's Butterflies*. London: Royal Collection Trust.

Holland, D., and D. Adams. (1998). "Puss Caterpillar" Envenomation: A Report from North Carolina. *Elsevier*, 10.1580/1080–6032.

Hossie, T. (2012). Possibly the Best Known Eyespot Caterpillar: *Hemeroplanes Sp.* (Sphingidae). Retrieved from http://caterpillar-eyespots.blogspot.ca/2012/01/possibly-best-known-eyespot-caterpillar.html.

Norris, J., Z. Carrim, and A. Morrell. (2010). Spiderman's Eye. *Lancet*, 10.1016/S0140-6736(09)61672-X.

Todd, K. (2007). *Chrysalis: Maria Sibylla Merian and the Secrets of Metamorphosis*. New York: Houghton Mifflin.

Zimmer, C. (2014). The Caterpillar Defense. *National Geographic*. Retrieved from http://phenomena.nationalgeographic.com/2014/12/10/the-caterpillar-defense/.

Ocean Sunfish

Crew, B. (2012). Get on Your Bike, *Phallostethus Cuulong*. *Scientific American*. Retrieved from http://blogs.scientificamerican.com/running-ponies/2012/07/25/get-on-your-bike-phallostethus-cuulong/.

Darwin, C. (2001). *The Voyage of the Beagle*. New York: Random House.

McGrouther, M. (2015). Ocean Sunfish, *Mola Mola* (Linnaeus, 1758). Australian Museum. Retrieved from http://australianmuseum.net.au/Ocean-Sunfish-Mola-mola.

Nakatsubo, T. (2008). *Mola Mola*. Retrieved from http://oceansunfish.org/NakatsuboDissertationsSum.pdf.

Simon, M. (2013). Absurd Creature of the Week: "Pufferfish on Steroids" Gets as Big as a Truck. *Wired*. Retrieved from http://www.wired.com/2013/12/absurd-creature-of-the-week-3/.

Thys, T. (2013). *Mola Mola*. Life History. Retrieved from http://www.oceansunfish.org/lifehistory.php.

Whale Shark: The World's Largest Fish. (2013). American Museum of Natural History. Retrieved from http://www.amnh.org/explore/news-blogs/on-exhibit-posts/whale-shark-the-world-s-largest-fish.

Lowland Streaked Tenrec

Eisenberg, J., and E. Gould. (1969). The Tenrecs: A Study in Mammalian Behavior and Evolution. *Smithsonian Contributions to Zoology*, 10.5479/si.00810282.27.

Marshall, C., and J. Eisenberg. (1996). *Hemicentetes Semispinosus*. *American Society of Mammalogists* 541: 1–4.

Quick Evolution Leads to Quiet Crickets. (2006). University of California, Berkeley. Retrieved from http://evolution.berkeley.edu/evolibrary/news/061201_quietcrickets.

Stephenson, P. (n.d.). Tenrecs in Madagascar. IUCN Afrotheria Specialist Group. Retrieved from www.afrotheria.net/tenrecs/.

Wallace's Explanation of Brilliant Colors in Caterpillar Larvae. (1867). Western Kentucky University. Retrieved from http://people.wku.edu/charles.smith/wallace/S129.htm.

Surinam Toad

Merian, M. S. (2010). *Insects of Surinam.* Cologne, Germany: Taschen.

Rabb, G., and M. Rabb. (1960). On the Mating and Egg-Laying Behavior of the Surinam Toad, *Pipa Pipa. American Society of Ichthyologists and Herpetologists,* 10.2307/1439751.

Roach, J. (2006). Grizzly-Polar Bear Hybrid Found—but What Does It Mean? *National Geographic.* Retrieved from http://news.nationalgeographic.com/news/2006/05/polar-bears.html.

Simon, M. (2013). Absurd Creature of the Week: The Frog Whose Young Erupt from Under Its Skin. *Wired.* Retrieved from http://www.wired.com/2013/12/absurd-creature-of-the-week-the-toad-whose-young-erupt-from-her-skin/.

Species. (n.d.). In Oxford Dictionaries online. Retrieved from http://www.oxforddictionaries.com/us/definition/american_english/species.

Wilkinson, M., E. Sherratt, F. Starace, and D. Gower. (2013). A New Species of Skin-Feeding Caecilian and the First Report of Reproductive Mode in Microcaecilia (Amphibia: Gymnophiona: Siphonopidae). *PLOS One,* 10.1371/journal.pone.0057756.

CHAPTER 3:
You Need a Place to Crash

Pearlfish

Crew, B. (2014). Here's How Pearlfish Call to Each Other from Inside the Bodies of Other Living Animals. *Scientific American.* Retrieved from http://blogs.scientificamerican.com/running-ponies/how-pearlfish-use-oysters-as-underwater-amplifiers-for-communication/.

Kéver, L., et al. (2014). Sound Production in *Onuxodon Fowleri* (Carapidae) and Its Amplification by the Host Shell. *Journal of Experimental Biology,* 10.1242/jeb.109363.

Lambert, P. (1997). *Sea Cucumbers of British Columbia, Southeast Alaska and Puget Sound.* Vancouver: Royal British Columbia Museum.

Mah, C. (2012). Sea Cucumber Evisceration! Defense! Regeneration! Why? Gross! *The Echinoblog.* Retrieved from http://echinoblog.blogspot.com/2012/01/sea-cucumber-evisceration-defense.html.

Parmentier, E., and P. Vandewalle. (2003). Morphological Adaptations of Pearlfish (Carapidae) to Their Various Habitats. *Science Publisher,* 261–76.

———. (2005). Further Insight on Carapid-Holothuroid Relationships. *Marine Biology* 146: 455–65.

Simon, M. (2014). Absurd Creature of the Week: This Fish Swims up a Sea Cucumber's Butt and Eats Its Gonads. *Wired.* Retrieved from http://www.wired.com/2014/02/absurd-creature-of-the-week-pearlfish/.

Tongue-Eating Isopod

Cook, C. (2012). The Early Life History and Reproductive Biology of *Cymothoa Excisa*, a Marine Isopod Parasitizing Atlantic Croaker (*Micropogonias Undulatus*), Along the Texas Coast. University of Texas at Austin. Retrieved from http://repositories.lib.utexas.edu/bitstream/handle/2152/ETD-UT-2012-08-6285/COOK-THESIS.pdf?sequence=1.

Krulwich, R. (2014). I Won't Eat, You Can't Make Me! (And They Couldn't). NPR. Retrieved from http://www.npr.org/blogs/krulwich/2014/02/22/280249001/i-wont-eat-you-cant-make-me-and-they-couldnt.

Lowry, J., and K. Dempsey. (2006). The Giant Deep-Sea Scavenger Genus *Bathynomus* (Crustacea, Isopoda, Cirolanidae) in the Indo-West Pacific. *Mémoires du Muséum National d'Histoire Naturelle* 193: 163–92.

Simon, M. (2013). Absurd Creature of the Week: This Parasite Eats a Fish's Tongue—and Takes Its Place. *Wired*. Retrieved from http://www.wired.com/2013/11/absurd-creature-of-the-week-the-parasite-that-eats-and-replaces-a-fishs-tongue/.

Zimmer, C. (2012). Tongue Parasites to People of Earth: Thank You for Your Overfishing. *National Geographic*. Retrieved from http://phenomena.nationalgeographic.com/2012/03/02/tongue-parasites-to-people-of-earth-thank-you-for-your-overfishing/.

Pistol Shrimp

Bernstein, J. (August 1, 1956). The Noisy Underwater World. *Milwaukee Journal*, 24.

Duffy, J. (n.d). Social Shrimp (Crustacea: Decapoda: Alpheidae: *Synalpheus*): Resources for Teaching. Virginia Institute of Marine Science. Retrieved from http://www.vims.edu/research/units/labgroups/marine_biodiversity/resources/Synalpheus%20teaching%20resources.pdf.

Duffy, J., and K. Macdonald. (1999). Colony Structure of the Social Snapping Shrimp *Synalpheus Filidigitus* in Belize. *Journal of Crustacean Biology*, 10.2307/1549235.

Eupectella Aspergillum (Venus' Flower Basket). (n.d.). Natural History Museum, London. Retrieved from http://www.nhm.ac.uk/nature-online/species-of-the-day/collections/our-collections/euplectella-aspergillum/uses/index.html (site discontinued).

Noise. (n.d.). American Speech-Language-Hearing Association. Retrieved from http://www.asha.org/public/hearing/Noise/.

Simon, M. (2014). Absurd Creature of the Week: The Feisty Shrimp That Kills with Bullets Made of Bubbles. *Wired*. Retrieved from http://www.wired.com/2014/07/absurd-creature-of-the-week-pistol-shrimp/.

Snapping Shrimp Drown Out Sonar with Bubble-Popping Trick, Described in *Science*. (2000). American Association for the Advancement of Science. Retrieved from http://www.sciencedaily.com/releases/2000/09/000922072104.htm.

Versluis, M., B. Schmitz, A. Von der Heydt, and D. Lohse. (2000). How Snapping Shrimp Snap: Through Cavitating Bubbles. *Science*, 10.1126/science.289.5487.2114.

Sociable Weaver

DelViscio, J. (2011). Housing Boom, If You're a Bird. *New York Times*. Retrieved from http://www.nytimes.com/2011/07/14/world/asia/14sukadana.html.

MacLean, C. (1972). *Island on the Edge of the World: The Story of St Kilda*. Edinburgh: Canongate Books.

Simon, M. (2014). Absurd Creature of the Week: The Bird That Builds Nests So Huge They Pull Down Trees. *Wired*. Retrieved from http://www.wired.com/2014/08/absurd-creature-of-the-week-the-bird-that-builds-nests-so-huge-they-pull-down-trees/.

Winch, P. (2007). Wildlife Spotlight: Northern Fulmar (*Fulmarus glacialis*). Farallones Marine Sanctuary Association. Retrieved from http://www.farallones.org/e_newsletter/2007-01/Fulmar.htm.

Hero Ant

Helms, J., C. Peeters, and B. Fisher. (2014). Funnels, Gas Exchange and Cliff Jumping: Natural History of the Cliff Dwelling Ant *Malagidris Sofina*. *Insectes Sociaux*, 10.1007/s00040-014-0360-8.

Simon, M. (2014). Absurd Creature of the Week: World's Most Badass Ant Skydives, Uses Own Head as a Shield. *Wired*. Retrieved from http://www.wired.com/2014/04/absurd-creature-of-the-week-the-amazing-skydiving-ant/.

———. (2015). Enter the Twilight Zone, Home to Earth's Strangest Reefs. *Wired*. Retrieved from http://www.wired.com/2015/05/twilight-zone-deep-reefs/.

CHAPTER 4:
You Live in a Crummy Neighborhood

Water Bear

Antony van Leeuwenhoek. (1996). University of California, Berkeley. Retrieved from http://www.ucmp.berkeley.edu/history/leeuwenhoek.html.

Bordenstein, S. (n.d.). Tardigrades (Water Bears). Carleton College. Retrieved from http://serc.carleton.edu/microbelife/topics/tardigrade/index.html.

Courtland, R. (2008). "Water Bears" Are First Animal to Survive Space Vacuum. *New Scientist*. Retrieved from http://www.newscientist.com/article/dn14690-water-bears-are-first-animal-to-survive-space-vacuum.html#.VK4D6qZj_Vo.

Dohrer, E. (2012). Laika the Dog and the First Animals in Space. *Space.com*. Retrieved from http://www.space.com/17764-laika-first-animals-in-space.html.

Goldberg, D. (2012). Why Can't We Get Down to Absolute Zero? *io9*. Retrieved from http://io9.com/5889074/why-cant-we-get-down-to-absolute-zero.

Goldstein, B., and M. Blaxter. (2002). Tardigrades. *Current Biology*, 10.1016/S0960-9822(02)00959-4.

Horikawa, D. (2008). The Tardigrade *Ramazzottius Varieornatus* as a Model Animal for Astrobiological Studies. *Biological Sciences in Space*, 10.2187/bss.22.93.

———. (2012). UV Radiation Tolerance of Tardigrades. NASA Astrobiology. Retrieved from https://astrobiology.nasa.gov/seminars/featured-seminar-channels/early-career-seminars/2012/04/24/uv-radiation-tolerance-of-tardigrades/.

Isachenkov, V. (2008). Russia Opens Monument to Space Dog Laika. *USA Today*. Retrieved from http://usatoday30.usatoday.com/news/world/2008-04-11-177105809_x.htm.

Jönsson, K., E. Rabbow, R. Schill, M. Harms-Ringdahl, and P. Rettberg. (2008). Tardigrades Survive Exposure to Space in Low Earth Orbit. *Current Biology*, 10.1016/j.cub.2008.06.04.

Simon, M. (2014). Absurd Creature of the Week: The Incredible Critter That's Tough Enough to Survive in Space. *Wired*. Retrieved from http://www.wired.com/2014/03/absurd-creature-week-water-bear/.

Surface Temperatures of the Inner Rocky Planets (2013). Earthguide and Scripps Institution of Oceanography. Retrieved from http://earthguide.ucsd.edu/eoc/special_topics/teach/sp_climate_change/p_planet_temp.html.

Wang, C., M. Grohme, B. Mali, R. Schill, and M. Frohme. (2014). Towards Decrypting Cryptobiosis—Analyzing Anhydrobiosis in the Tardigrade *Milnesium Tardigradum* Using Transcriptome Sequencing. *PLOS One*, 10.1371/journal.pone.0092663.

Diving Bell Spider

McCook, H. (1890). *American Spiders and Their Spinningwork. A Natural History of the Orbweaving Spiders of the United States, with Special Regard to Their Industry and Habits*. Philadelphia: Academy of Natural Science of Philadelphia.

Schütz, D., and M. Taborsky. (2003). Adaptations to an Aquatic Life May Be Responsible for the Reversed Sexual Size Dimorphism in the Water Spider, *Argyroneta Aquatica*. *Evolutionary Ecology Research* 5: 105–17.

Seymour, R., and S. Hetz. (2011). The Diving Bell and the Spider: The Physical Gill of *Argyroneta Aquatica*. *Journal of Experimental Biology*, 10.1242/jeb.056093.

Simon, M. (2014). Absurd Creature of the Week: The Incredible Spider That Lives Its Entire Life Underwater. *Wired*. Retrieved from http://www.wired.com/2014/09/absurd-creature-week-incredible-spider-lives-entire-life-underwater/.

————. (2015). Absurd Creature of the Week: The Insect That Devours Its Mother from the Inside Out. *Wired*. Retrieved from http://www.wired .com/2015/01/absurd-creature-of-the-week-strepsiptera/.

Zombie Ant

Currie, C., J. Scott, R. Summerbell, and D. Malloch. (1999). Fungus-Growing Ants Use Antibiotic-Producing Bacteria to Control Garden Parasites. *Nature*, 10.1038/19519.

Hughes, D., T. Wappler, and C. Labandeira. (2010). Ancient Death-Grip Leaf Scars Reveal Ant-Fungal Parasitism. *Biology Letters*, 10.1098/rsbl.2010.0521.

Simon, M. (2013). Absurd Creature of the Week: The Zombie Ant and the Fungus That Controls Its Mind. *Wired*. Retrieved from http://www.wired .com/2013/09/absurd-creature-of-the-week-zombie-ant-fungus/.

Pink Fairy Armadillo

Simon, M. (2014). Absurd Creature of the Week: The Adorable Mexican Mole Lizard Has a Disgusting Reputation. *Wired*. Retrieved from http://www .wired.com/2014/12/absurd-creature-of-the-week-mexican-mole-lizard/.

————. (2014). Absurd Creature of the Week: Pink Fairy Armadillo Crawls out of the Desert and into Your Heart. *Wired*. Retrieved from http://www .wired.com/2014/01/absurd-creature-of-the-week-pink-fairy-armadillo -crawls-out-of-the-desert-and-into-our-hearts/.

Superina, M. (2011). Husbandry of a Pink Fairy Armadillo (*Chlamyphorus Truncatus*): Case Study of a Cryptic and Little Known Species in Captivity. *Zoo Biology*, 10.1002/zoo.20334.

Naked Mole Rat

Dayton, P., M. Graham, and J. Parker, eds. (2011). *The Essential Naturalist*. Chicago: University of Chicago Press.

First Person: Stephen Jay Gould on Evolution. (1995). Voyager Company. Retrieved from https://www.youtube.com/watch?v=v0BhXVLKIz8.

Naked Mole Rat. (n.d.). Louisville Zoo. Retrieved from http://www .louisvillezoo.org/animals/MammalFS/Naked-Mole-Rat.pdf (site discontinued).

Simon, M. (2014). Absurd Creature of the Week: The Naked Mole Rat Could One Day Save Your Life. *Wired*. Retrieved from http://www.wired.com/2014/09/ absurd-creature-of-the-week-naked-mole-rat/.

UIC Research: The Life and Times of Naked Mole-Rats. (2012). University of Illinois at Chicago. Retrieved from https://www.youtube.com/ watch?v=jHm0jmg-sbc.

Yahav, S., and R. Buffenstein. (1991). Huddling Behavior Facilitates Homeothermy in the Naked Mole Rat *Heterocephalus Glaber*. *Physiological Zoology*, 10.2307/30158212.

CHAPTER 5:
Turns Out Getting Eaten Is Bad for Survival

Hagfish

Bernards, M., I. Oke, A. Heyland, and D. Fudge. (2014). Spontaneous Unraveling of Hagfish Slime Thread Skeins Is Mediated by a Seawater-Soluble Protein Adhesive. *Journal of Experimental Biology*, 10.1242/jeb.096909.

Fish Dissection—Gills Exposed. (2012). Australian Museum. Retrieved from http://australianmuseum.net.au/image/Fish-Dissection-Gills-exposed/.

Lim, J., D. Fudge, N. Levy, and J. Gosline. (2006). Hagfish Slime Ecomechanics: Testing the Gill-Clogging Hypothesis. *Journal of Experimental Biology*, 10.1242/jeb.02067.

Simon, M. (2014). Absurd Creature of the Week: This Oceanic "Nightmare" Suffocates Foes with Clouds of Slime. *Wired*. Retrieved from http://www.wired.com/2014/05/absurd-creature-of-the-week-hagfish/.

Sharks and Rays: Myth and Reality. (2001). American Museum of Natural History. Retrieved from http://www.amnh.org/learn/pd/sharks_rays/rfl_myth/index.html.

Axolotl

Simon, M. (2014). Absurd Creature of the Week: The Real-Life Pokémon That Can Regenerate Missing Limbs. *Wired*. Retrieved from http://www.wired.com/2014/04/absurd-creature-of-the-week-the-adorable-salamander-that-can-regrow-amputated-limbs/.

Cuttlefish

Barbosa, A., et al. (2008). Cuttlefish Camouflage: The Effects of Substrate Contrast and Size in Evoking Uniform, Mottle or Disruptive Body Patterns. *Vision Research*, 10.1016/j.visres.2008.02.011.

Deravi, L., et al. (2014). The Structure–Function Relationships of a Natural Nanoscale Photonic Device in Cuttlefish Chromatophores. *Royal Society Publishing, Interface*, 10.1098/rsif.2013.0942.

Kröger, B., J. Vinther, and D. Fuchs. (2011). Cephalopod Origin and Evolution: A Congruent Picture Emerging from Fossils, Development and Molecules. *Bioessays*, 0.1002/bies.201100001.

McClain, C., et al. (2015). Sizing Ocean Giants: Patterns of Intraspecific Size Variation in Marine Megafauna. *PeerJ*, 10.7717/peerj.715.

Ramirez, M., and T. Oakley. (2015). Eye-Independent, Light-Activated Chromatophore Expansion (LACE) and Expression of Phototransduction Genes in the Skin of *Octopus Bimaculoides*. *Journal of Experimental Biology*, 10.1242/jeb.110908.

Simon, M. (2014). Absurd Creature of the Week: Cross-Dressing Cuttlefish Puts on World's Most Spectacular Light Show. *Wired*. Retrieved from http://www.wired.com/2014/04/absurd-creature-of-the-week-cuttlefish/.

Vendetti, J. (2006). The Cephalopoda. University of California, Berkeley. Retrieved from http://www.ucmp.berkeley.edu/taxa/inverts/mollusca/cephalopoda.php.

Wood, J., and K. Jackson. (2004). How Cephalopods Change Color. The Cephalopod Page. Retrieved from http://www.thecephalopodpage.org/cephschool/HowCephalopodsChangeColor.pdf.

Satanic Leaf-Tailed Gecko

Cook, L., B. Grant, I. Saccheri, and J. Mallet. (2012). Selective Bird Predation on the Peppered Moth: The Last Experiment of Michael Majerus. *Royal Society Publishing Biology Letters*, 10.1098/rsbl.2011.1136.

Gilbert, J. (2015). Secrets of the Orchid Mantis Revealed—It Doesn't Mimic an Orchid After All. *The Conversation*. Retrieved from http://theconversation.com/secrets-of-the-orchid-mantis-revealed-it-doesnt-mimic-an-orchid-after-all-36715.

Miller, K. (1999). The Peppered Moth—an Update. Retrieved from http://www.millerandlevine.com/km/evol/Moths/moths.html.

Simon, M. (2014). Absurd Creature of the Week: Satanic Leaf-Tailed Gecko Wears the World's Most Unbelievable Camo. *Wired*. Retrieved from http://www.wired.com/2014/07/absurd-creature-of-the-week-satanic-leaf-tailed-gecko/.

Where Did All of Madagascar's Species Come From? (2009). University of California, Berkeley. Retrieved from http://evolution.berkeley.edu/evolibrary/news/091001_madagascar.

Pangolin

Anchors, J. (2002). Zimbabwean Land and Zimbabwean People: Creative Explorations. University of Maine Electronic Theses and Dissertations, Paper 41.

Beebe, W. (1914). The Pangolin or Scaly Anteater. *Zoological Society Bulletin* 17: 1141–45.

Challender, D., and L. Hywood. (2012). African Pangolins Under Increased Pressure from Poaching and Intercontinental Trade. *Traffic Bulletin* 24(2).

Chinese Medicine and the Pangolin. (1938). *Nature*, 10.1038/141072b0.

Darwin, C. (2001). *The Voyage of the Beagle*. New York: Random House.

Eating Pangolins to Extinction. (2014). IUCN. Retrieved from http://www.iucn.org/news_homepage/?17189/Eating-pangolins-to-extinction.

Fasman, J. (2014). Taste for Rare, Wild Pangolin Is Driving the Mammal to Extinction. NPR. Retrieved from http://www.npr.org/blogs/thesalt/2014/08/03/337162283/ taste- for-rare-wild-pangolin-is-driving-the-mammal-to-extinction.

Miller, R., and M. Fowler, eds. (2015). *Fowler's Zoo and Wild Animal Medicine*. St. Louis: Elsevier Saunders.

Pantel, S., and S. Chin. (2009). *Proceedings of the Workshop on Trade and*

Conservation of Pangolins Native to South and Southeast Asia. Petaling Jaya: Traffic Southeast Asia.

Quammen, D. (February 2009). Darwin's First Clues. *National Geographic*

Soewu, D., and O. Sodeinde. (2015). Utilization of Pangolins in Africa: Fuelling Factors, Diversity of Uses and Sustainability. *International Journal of Biodiversity and Conservation*, 10.5897/IJBC2014.0760.

Crested Rat

Bubenik, G. (2003). Why Do Humans Get "Goosebumps" When They Are Cold, or Under Other Circumstances? *Scientific American*. Retrieved from http://www.scientificamerican.com/article/why-do-humans-get-goosebu/.

Kingdon, J., et al. (2012). A Poisonous Surprise Under the Coat of the African Crested Rat. *Proceedings of the Royal Society B*, 10.1098/rspb.2011.1169.

Neuwinger, H. (1994). *African Ethnobotany: Poisons and Drugs: Chemistry, Pharmacology, Toxicology*. Stuttgart, Germany: Chapman and Hall.

Simon, M. (2014). Absurd Creature of the Week: This Crazy-Looking Sea Slug Has an Ingenious Secret Weapon. *Wired*. Retrieved from http://www.wired.com/2014/11/absurd-creature-week-nudibranch-sea-slug/.

CHAPTER 6:
It Turns Out Not Eating Is Also Bad for Survival

Giant African Land Snail

Bates, M. (2014). Faced with Invasive Snails, a Bird Learns to Use Tools. *Wired*. Retrieved from http://www.wired.com/2014/11/faced-invasive-snail-bird-learns-use-tools/.

Giant African Land Snail (GALS) aka Giant African Snail (GAS) (Liss) *Achatina Fulica* (Férussac 1821). (n.d.). University of Florida. Retrieved from http://miami-dade.ifas.ufl.edu/pdfs/ornamental/GALS%20ID.pdf.

Giant African Land Snails Fact Sheet. (n.d.). United States Department of Agriculture, Animal and Plant Health Inspection Service. Retrieved from http://www.michigan.gov/documents/MDA_Giant_African_Land_Snail_Fact_Sheet_92709_7.pdf.

Kleeman, S. (2014). 49 Crazy Headlines That Could Only Be Created by Florida Man. *Mic News*. Retrieved from http://mic.com/articles/107372/49-tremendous-things-florida-men-accomplished-this-year.

Koplowitz, H. (2014). "Fake Cop" Matt Skytta, Florida Man, Shows IHOP Server His Butt When Free Food Request Doesn't Work: Police. *International Business Times*. Retrieved from http://www.ibtimes.com/fake-cop-matt-skytta-florida-man-shows-ihop-server-his-butt-when-free-food-request-doesnt-work.

Ovalle, D. (2010). Giant African Snails Smuggled Into Florida for Use in Religious Ritual, Authorities Say. *Miami Herald*. Retrieved from http://

articles.sun-sentinel.com/2010-03-11/news/fl-illegal-snails
-santeria-20100310_1_snails-smuggled-search-warrant.

Simon, M. (2014). Absurd Creature of the Week: Foot-Long, Sex-Crazed Snails That Pierce Tires and Devour Houses. *Wired*. Retrieved from http://www.wired.com/2014/01/absurd-creature-of-the-week-foot-giant-african-land-snail/.

Smith, T., L. Whilby, and A. Derksen. (2010). 2010 Florida CAPS Giant African Snail Survey Report. Florida Cooperative Agricultural Pest Survey. Retrieved from http://freshfromflorida.s3.amazonaws.com/pdf_2010_giant_african_snail_survey_report_03-11-2010.pdf.

Varona, E. (2012). Escargot? More like Escar-No! USDA APHIS. Retrieved from http://blogs.usda.gov/2012/04/19/escargot-more-like-escar-no/.

Wright, C. (2014). Deputies: Drunk Man Called 911 to See if Tax Return Had Come In. *Tampa Bay Times*. Retrieved from http://www.tampabay.com/news/publicsafety/crime/deputies-drunk-man-called-911-to-see-if-tax-return-had-come-in/2164927.

Aye-Aye

Darwin, C. (1958). *The Autobiography of Charles Darwin*. New York: W. W. Norton.

Gould, S. (2002). *The Structure of Evolutionary Theory*. Cambridge, MA: Harvard University Press.

Gross, C. (1993). Hippocampus Minor and Man's Place in Nature: A Case Study in the Social Construction of Neuroanatomy. *Hippocampus* 3(4): 403–15.

Owen, R. (1863). *Monograph on the Aye-Aye* (Chiromys Madagascariensis, Cuvier). London: Taylor and Francis.

Simon, M. (2013). Absurd Creature of the Week: Aye-Aye Gives World the Highly Elongated Finger. *Wired*. Retrieved from http://www.wired.com/2013/09/absurd-creature-of-the-week-aye-aye-gives-world-the-highly-elongated-finger/.

———. (2015). Absurd Creature of the Week: The Tiny Primate That Was Probably the Inspiration for Yoda. *Wired*. Retrieved from http://www.wired.com/2015/01/absurd-creature-of-the-week-tarsier/.

Switek, B. (2008). Richard Owen, the Forgotten Evolutionist. *Science Blogs*. Retrieved from http://scienceblogs.com/laelaps/2008/10/09/richard-owen-the-forgotten-evo/.

Mantis Shrimp

Patek, S. (2004). The Shrimp with a Kick! TED Conferences. Retrieved from http://www.ted.com/talks/sheila_patek_clocks_the_fastest_animals.

Simon, M. (2013). Absurd Creature of the Week: Enormous Hermit Crab Tears Through Coconuts, Eats Kittens. *Wired*. Retrieved from http://www.wired.com/2013/12/absurd-creature-of-the-week-2/.

———. (2014). Absurd Creature of the Week: Ferocious, Fearless Mantis Shrimp

Is the Honey Badger of the Sea. *Wired*. Retrieved from http://www.wired.com/2014/01/absurd-creature-of-the-week-4/.

———. (2015). Absurd Creature of the Week: The Extraordinary Light Show of the Disco Clam. *Wired*. Retrieved from http://www.wired.com/2015/05/absurd-creature-of-the-week-disco-clam/.

Bone-Eating Worm

Barry, K., G. Holwell, and M. Herberstein. (2008). Female Praying Mantids Use Sexual Cannibalism as a Foraging Strategy to Increase Fecundity. *Behavioral Ecology*, 10.1093/beheco/arm156.

Danise, S., and N. Higgs. (2015). Bone-Eating Osedax Worms Lived on Mesozoic Marine Reptile Deadfalls. *Royal Society Publishing Biology Letters*, 10.1098/rsbl.2015.0072.

Goffredi, S., et al. (2005). Evolutionary Innovation: A Bone-Eating Marine Symbiosis. *Environmental Microbiology*, 10.1111/j.1462-2920.2005.00824.x.

Rouse, G., et al. (2015). A Dwarf Male Reversal in Bone-Eating Worms, *Current Biology*, 10.1016/j.cub.2014.11.032.

Rouse, G., S. Goffredi, and R. Vrijenhoek. (2004). Osedax: Bone-Eating Marine Worms with Dwarf Males. *Science*, 10.1126/science.1098650.

Tresguerres, M., S. Katz, and G. Rouse. (2013). How to Get Into Bones: Proton Pump and Carbonic Anhydrase in Osedax Boneworms. *Proceedings of the Royal Society B*, 10.1098/rspb.2013.0625.

Wilson, E. O. (1999). *The Diversity of Life*. New York: W. W. Norton.

Tiger Beetle

Bouchard, P., ed. (2014). *The Book of Beetles: A Life-Size Guide to Six Hundred of Nature's Gems*. Chicago: University of Chicago Press.

Friedlander, B. (1998). When Tiger Beetles Chase Prey at High Speeds They Go Blind Temporarily, Cornell Entomologists Learn. *Cornell Chronicle*. Retrieved from http://www.news.cornell.edu/stories/1998/01/tiger-beetles-go-blind-chasing-prey-high-speeds.

Lomakin, J., et al. (2011). Mechanical Properties of the Beetle Elytron, a Biological Composite Material. *Biomacromolecules*, 10.1021/bm1009156.

Pearson, D. (2011). Six-Legged Tigers. *Wings: Essays on Invertebrate Conservation*. Retrieved from http://www.xerces.org/wp-content/uploads/2008/06/Wings_sp11_tiger-beetles.pdf.

Pearson, D., and A. Vogler. (2001). Tiger Beetles: The Evolution, Ecology, and Diversity of the Cicindelids. *Zoosystematics and Evolution*, 10.1002/mmnz.20040800126.

Walks and Treks FAQs. (n.d.). British Heart Foundation. Retrieved from https://www.bhf.org.uk/get-involved/events/training-zone/walking-training-zone/walking-faqs.

Wallace, A. (2000). *The Malay Archipelago*. North Clarendon, VT: Periplus Editions.

Zurek, D., and C. Gilbert. (2014). Static Antennae Act as Locomotory Guides That Compensate for Visual Motion Blur in a Diurnal, Keen-Eyed Predator. *Proceedings of the Royal Society B*, 10.1098/rspb.2013.3072.

CHAPTER 7:
You Can't Let Them Get Away That Easily, Can You?

Bolas Spider

Eberhard, W. (1977). Aggressive Chemical Mimicry by a Bolas Spider. *Science*, 10.1126/science.198.4322.1173.

———. (1980). The Natural History and Behavior of the Bolas Spider *Mastophora Dizzydeani* (Araneidae). *Psyche: A Journal of Entomology*, 10.1155/1980/81062.

Haynes, K., et al. (2002). Aggressive Chemical Mimicry of Moth Pheromones by a Bolas Spider: How Does This Specialist Predator Attract More Than One Species of Prey? *Chemoecology*, 10.1007/s00049-002-8332-2.

Haynes, K., K. Yeargan, and C. Gemeno. (2001). Detection of Prey by a Spider That Aggressively Mimics Pheromone Blends. *Journal of Insect Behavior*, 10.1023/A:1011128223782.

Liu, M., S. Blamires, C. Liao, and I. Tso. (2014). Evidence of Bird Dropping Masquerading by a Spider to Avoid Predators. *Scientific Reports*, 10.1038/srep05058.

Scharff, N., and G. Hormiga. (2012). First Evidence of Aggressive Chemical Mimicry in the Malagasy Orb Weaving Spider *Exechocentrus Lancearius* Simon, 1889 (Arachnida: Araneae: Araneidae) and Description of a Second Species in the Genus. *Arthropod Systematics and Phylogeny* 70(2), 107–18.

Stringer, I. (1967). The Larval Behaviour of the New Zealand Glow-Worm *Arachnocampa Luminosa. Tane* 13: 107–17.

Yeargan, K. (1994). Biology of Bolas Spiders. *Annual Review of Entomology*, 10.1146/annurev.en.39.010194.000501.

Velvet Worm

Animal Species: Velvet Worm. (2010). Australian Museum. Retrieved from http://australianmuseum.net.au/velvet-worm.

Bhatia, A. (2013). The Fluid Dynamics of Spitting: How Archerfish Use Physics to Hunt with Their Spit. *Wired*. Retrieved from http://www.wired.com/2013/11/archerfish-physics/.

Concha, A., et al. (2015). Oscillation of the Velvet Worm Slime Jet by Passive Hydrodynamic Instability. *Nature Communications*, 10.1038/ncomms7292.

Mayer, G., et al. (2015). Capture of Prey, Feeding, and Functional Anatomy of the Jaws in Velvet Worms (Onychophora). *Integrative and Comparative Biology*, 10.1093/icb/icv004.

Ogg, B. (2008). Managing Centipedes and Millipedes. University of Nebraska,

Lincoln. Retrieved from http://lancaster.unl.edu/pest/resources/
CentipedeMillipede012.shtml.

Read, V., and R. Hughes. (1987). Feeding Behaviour and Prey Choice in
Macroperipatus torquatus (Onychophora). *Proceedings of the Royal Society
B*, 10.1098/rspb.1987.0030.

Simon, M. (2014). Absurd Creature of the Week: Voracious Velvet Worm
Ensnares Foes with Jets of Slime. *Wired*. Retrieved from http://www.wired
.com/2014/08/absurd-creature-of-the-week-velvet-worm/.

Woo, M. (2015). How the Velvet Worm Pulls Off Its Bizarre Slime Attack. *Wired*.
Retrieved from http://www.wired.com/2015/03/velvet-worm-pulls-off
-bizarre-slime-attack/.

Geography Cone Snail

Chadwick, A. (2015). Exploring Cone Snails and Science. Olivera Lab,
University of Utah. Retrieved from http://www.theconesnail.com/.

Nemy, E. (2008). Sunny von Bülow, 76, Focus of Society Drama, Dies. *New York
Times*. Retrieved from http://www.nytimes.com/2008/12/07/
nyregion/07vonbulow.html?pagewanted=all&_r=0.

Olivera, B. (2009). Biodiversity at a Snail's Pace. Howard Hughes Medical
Institute Holiday Lectures. Retrieved from http://media.hhmi.org/
hl/09Lect3.html.

Safavi-Hemami, H., et al. (2015). Specialized Insulin Is Used for Chemical
Warfare by Fish-Hunting Cone Snails. *PNAS*, 10.1073/pnas.1423857112.

Sarramegna, R. (1965). Poisonous Gastropods of the Conidae Family Found in
New Caledonia and the Indo-Pacific. South Pacific Commission Technical
Paper 144.

Simon, M. (2014). Absurd Creature of the Week: The Parasitic Worm That Turns
Snails Into Disco Zombies. *Wired*. Retrieved from http://www.wired
.com/2014/09/absurd-creature-of-the-week-disco-worm/.

Lamprey

Chillag, I. (2012). A Parasite Pie Fit for a Queen's Diamond Jubilee. NPR.
Retrieved from http://www.npr.org/sections/waitwait/2012/06/03
/154196783/a-parasite-fit-for-a-queen.

Funny Fish Falling from the Sky! (2015). Alaska Department of Fish and Game.
Retrieved from https://www.facebook.com/media/set/?se
t=a.856880884378885.1073741851.322167891183523.

Gess, R., M. Coates, and B. Rubidge. (2006). A Lamprey from the Devonian
Period of South Africa. *Nature*, doi:10.1038/nature05150.

Gill, H., et al. (2003). Phylogeny of Living Parasitic Lampreys
(Petromyzontiformes) Based on Morphological Data. *Copeia*, 10.1643/IA02–
085.1.

History of Life on Earth. BBC. Retrieved from http://www.bbc.co.uk/nature/
history_of_the_earth.

Izadi, E. (2015). Why These Mysterious, Blood-Sucking Fish Fell from the Alaskan Sky. *Washington Post*. Retrieved from http://www.washingtonpost .com/news/speaking-of-science/wp/2015/06/12/why-these-mysterious -blood-sucking-fish-fell-from-the-alaskan-sky/.

Kircheis, F. W. (2004). Sea Lamprey. Retrieved from http://www.fws.gov/ GOMCP/pdfs/lampreyreport.pdf.

Renaud, C. (2011). Lampreys of the World: An Annotated and Illustrated Catalogue of Lamprey Species Known to Date. Food and Agriculture Organization of the United Nations. FAO Species Catalogue for Fishery Purposes No. 5.

Sea Lamprey: The Battle Continues. (1998). Great Lakes Fishery Commission. Retrieved from http://www.seagrant.umn.edu/ais/sealamprey_battle.

Sea Lamprey: A Great Lakes Invader. Great Lakes Fishery Commission. Retrieved from http://www.glfc.org/sealamp/.

Shimeld, S., and P. Donoghue. (2012). Evolutionary Crossroads in Developmental Biology: Cyclostomes (Lamprey and Hagfish). *Development*, 10.1242/ dev.074716.

Simon, M. (2014). Absurd Creature of the Week: The Aquatic Menace That Gives the Worst Hickeys Ever. *Wired*. Retrieved from http://www.wired .com/2014/07/absurd-creature-of-the-week-lamprey/.

Assassin Bug

Bulbert, M., M. Herberstein, and G. Cassis. (2014). Assassin Bug Requires Dangerous Ant Prey to Bite First. *Current Biology*, 10.1016/j.cub.2014.02.006.

Chagas Disease (American Trypanosomiasis). (2015). World Health Organization. Retrieved from http://www.who.int/mediacentre/ factsheets/fs340/en/.

McCalman, I. (2009). *Darwin's Armada: Four Voyages and the Battle for the Theory of Evolution*. New York: W. W. Norton.

Orrego, F., and Quintana, C. (2007). Darwin's Illness: A Final Diagnosis. *Royal Society Notes and Letters*, 10.1098/rsnr.2006.0160.

Simon, M. (2014). Absurd Creature of the Week: The Ferocious Bug That Sucks Prey Dry and Wears Their Corpses. *Wired*. Retrieved from http://www .wired.com/2014/06/absurd-creature-of-the-week-assassin-bug/.

Wallace, A. (1852). Letter concerning the Fire on the "Helen." Retrieved from http://people.wku.edu/charles.smith/wallace/S007.htm.

Wallace Is Shipwrecked and Loses His Collections. Natural History Museum, London. Retrieved from http://www.nhm.ac.uk/nature-online/collections -at-the-museum/wallace-collection/item.jsp?itemID=59 (site discontinued).